Bi & Gi Publishers

Current Topics in Rehabilitation
Series Editor : R. Corsico M.D.

Titles in the series:

Respiratory Muscles in Chronic Obstructive Pulmonary Disease
Edited by: A. Grassino, C. Fracchia, C. Rampulla, L. Zocchi

Pathophysiology and Treatment of Pulmonary Circulation
Edited by: A. Morpurgo, R. Tramarin, C. Rampulla, C. Fracchia, F. Cobelli

Chronic Pulmonary Hyperinflation
Edited by: A. Grassino, C. Rampulla, N. Ambrosino, C. Fracchia

Nutrition and Ventilatory Function
Edited by: R.D. Ferranti, C. Rampulla, C. Fracchia, N. Ambrosino

Right Ventricular Hypertrophy and Function in Chronic Lung Disease
Edited by: V. Ježeck, M. Morpurgo, R. Tramarin

Forthcoming titles in the series:

Pulmonary Rehabilitation
Edited by: A. Grassino, C. Fracchia, C. Rampulla

Acknowledgments: The Organizing Committee of the Workshop *"Update in Biochemistry of Pulmonary Emphysema"* held in Pavia (Italy) September 7, 1990, gratefully acknowledges the following Institutions for their interest and support:
- Università degli Studi di Pavia, Italy
- Policlinico San Matteo, Istituto di Ricovero e cura a Carattere Scientifico, Pavia, Italy
- Fondazione Rhône-Poulenc Rorer per le Scienze Mediche, Origgio (VA), Italy
- Fondazione S. Maugeri, Pavia, Italy

Biochemistry of Pulmonary Emphysema

Edited by: C. Grassi, J. Travis
L. Casali, M. Luisetti

Foreword by R. Corsico
With 41 figures and 26 tables

Springer-Verlag London Ltd.

C. Grassi	Institute of Phthisiology and Respiratory Diseases, IRCCS San Matteo, University of Pavia, Italy
J. Travis	Department of Biochemistry, University of Georgia, Athens, GA, USA
L. Casali	Postgraduate School of Respiratory Diseases, University of Pavia, Italy
M. Luisetti	Institute of Phthisiology and Respiratory Diseases, IRCCS San Matteo, University of Pavia, Italy

Series Editor:
R. Corsico   Clinica del Lavoro Foundation, Institute of Care and Research, Medical Center of Rehabilitation, Montescano (Pavia), Italy

Distribution

Sole distribution rights outside Italy granted to Springer-Verlag.
All orders for this title should be sent to following addresses:
Italy
Bi & Gi Publishers - Via Ca' di Cozzi, 41, 37124 Verona, Italy
North America
Springer-Verlag New York Inc. - 175 Fifth Avenue, New York, NY 10010, USA
Japan
Springer-Verlag - 37-33 Hongo 3-chome, Bunkyo-Ku, Tokyo 113, Japan
Rest of the world
Springer-Verlag - Heidelberger Platz 3, 1000 Berlin 33, Germany

ISBN 978-1-4471-3773-3

British Library Cataloguing in Publication Data
Biochemistry of Pulmonary Emphysema
(Current Topics in Rehabilitation Series)
I. Grassi, Carlo II. Series 616.2

Library of Congress Cataloging-in-Publication Data
Biochemistry of pulmonary emphysema/edited by C. Grassi ... (et al.): foreword by R.Corsico
(Current Topics in Rehabilitation)
Workshop "Update in Biochemistry of Pulmonary Emphysema" held in Pavia (Italy) September 7, 1990 -- P. preceding t.p. Includes index
ISBN 978-1-4471-3773-3   ISBN 978-1-4471-3771-9 (eBook)
DOI 10.1007/978-1-4471-3771-9

1. Emphysema, Pulmonary --Pathophysiology--Congresses, 2. Clinical biochemistry -- Congresses. 3. Proteinase--Pathophysiology--Congresses. 4. Proteinase--Inhibitors--Pathophysiology--Congresses.
I. Grassi, Carlo 1926- . II. Workshop "Update in Biochemistry of Pulmonary Emphysema" 1990: Pavia, Italy
III. Series.(DNLM: 1. Protease Inhibitors--metabolism--congresses.
2. Pulmonary Emphysema--congresses. WF 648 B6148 1990)
RC776.E5B55 1992 / 616.2'4807--dc20
DNLM/DLC - for Library of Congress 92-2315 CIP

Preface

Pulmonary emphysema is a disease which develops because of a localized imbalance between endogenous proteinase inhibitors and proteinases leaking from neutrophils during phagocytosis at inflammatory foci within the lung.

This volume not only reviews at a biochemical level what is known about the natural inhibitors and proteinases involved in connective tissue destruction within the lung, but also suggests novel methodologies for reestablishing proper enzyme-inhibitor balance, including the use of natural or synthetic inhibitors for supplementation or gene therapy.

The organizers of this meeting are pleased to dedicate this volume to the memories of those who have made important contributions to our understanding of pulmonary emphysema, especially Dr. Aaron Janoff and Dr. Philip Kimbel.

James Travis
Carlo Grassi

Foreword

It could happen that many readers of this volume are surprised at the inclusion of such a topic (i.e. the biochemistry of pulmonary emphysema) in a Series devoted to rehabilitation.

Nevertheless, I think that covering the biochemical events leading to the onset of pulmonary emphysema and, consequently, the respiratory failure linked to the chronic and disabling airflow limitation, represents the continuation of a talk we started and developed in the past few years. As a matter of fact, we previously reviewed different aspects of respiratory failure: muscles, pulmonary circulation, various aspects linked to chronic pulmonary hyperinflation, nutrition and ventilatory function. By means of the topics gathered in this volume, we aimed at reviewing biochemical mechanisms and related therapeutic strategies of chronic obstructive pulmonary disease.

Another reason prompted the Editors of this volume. On May 16-18, 1990 a wonderful conference entitled "Pulmonary Emphysema: the Rationale for Therapeutic Intervention", sponsored by The New York Academy of Sciences and the American Thoracic Society, was held in Lake Buena Vista, FL. It celebrated the knowledge matured during 25 years of research on pulmonary emphysema. Since one of the two key papers, representing the starting point from which the protease/ protease inhibitor imbalance hypothesis has developed, was published by Swedish investigators, it seemed right to the organizers of the workshop "Update in Biochemistry of Pulmonary Emphysema" , held in Pavia, Italy, on September 9, 1990, to celebrate that anniversary even on the other side of the Atlantic Ocean.

I would like to thank all the contributors of this volume for the high standard of their papers, that give us the chance to offer the readers an excellent updating concerning this exciting field.

April 1992

RENATO CORSICO

Contents

X

Contributors

BAICI A.
Department of Rheumatology, University Hospital, Zurich, Switzerland

BANGALORE N.
Department of Biochemistry, University of Georgia, Athens, Georgia, USA

BARRETT A.J.
Department of Biochemistry, Strangeways Research Laboratory, Cambridge, UK

BIETH J.G.
INSERM Unit 237, Université Louis Pasteur de Strasbourg, France

BIRRER P.
Pulmonary Branch, National Heart, Lung and Blood Institute, National Institutes of Health, Bethesda, Maryland, USA

BURNETT D.
Lung Immunobiochemical Research Laboratory, The General Hospital, Birmingham, UK

CHANG-STROMAN L.M.
Pulmonary Branch, National Heart, Lung and Blood Institute, National Institutes of Health, Bethesda, Maryland, USA

COLLINS J.
Gesellschaft für Bioteknologische Forschung, Hannover, Germany

CRYSTAL R.G.
Pulmonary Branch, National Heart, Lung and Blood Institute, National Institutes of Health, Bethesda, Maryland, USA

DAVIDSON J.M.
Research Service, Department of Veterans Affairs Medical Center and Department of Pathology, Vanderbilt University School of Medicine, Nashville, Tennessee, USA

DIJKMAN J. H.
Department of Pulmonology, University Hospital of Leiden, The Netherlands

FRITZ H.
Department of Clinical Chemistry and Clinical Biochemistry, University of Munich, Germany

GRASSI C.
Institute of Phthisiology and Respiratory Diseases, IRCCS San Matteo, University of Pavia, Italy

HORI H.
School of Chemistry, Georgia Institute of Technology, Atlanta, Georgia, USA

JOCHUM M.
Department of Clinical Chemistry and Clinical Biochemistry, University of Munich, Germany

KAM C.-M.
School of Chemistry, Georgia Institute of Technology, Atlanta, Georgia, USA

KRAMPS J.A.
Department of Pulmonology, University Hospital of Leiden, The Netherlands

KURDOWSKA A.
Institute for Molecular Biology, Jagiellonian University, Krakow, Poland

LUCEY E.C.
Boston University and Tufts University Schools of Medicine, Boston, Massachusetts, USA

LUISETTI M.
Institute of Phthisiology and Respiratory Diseases, IRCCS San Matteo, University of Pavia, Italy

McELVANEY N.G.
Pulmonary Branch, National Heart, Lung and Blood Institute, National Institutes of Health, Bethesda, Maryland, USA

MEYER E.F. JR.
Department of Biochemistry, Texas A and M University, College Station, Texas, USA

OLEKSYSZYN J.
School of Chemistry, Georgia Institute of Technology, Atlanta, Georgia, USA

POTEMPA J.
Institute for Molecular Biology, Jagiellonian University, Krakow, Poland

PONDER K.
Department of Cell Biology, Baylor College of Medicine, Houston, Texas, USA

POWERS J.C.
School of Chemistry, Georgia Institute of Technology, Atlanta, Georgia, USA

RUDOLPHUS A.
Department of Pulmonology, University Hospital of Leiden, The Netherlands

SIFERS R.N.
Department of Cell Biology and Department of Pathology, Baylor College of Medicine, Houston, Texas, USA

SNIDER G.L.
The Boston Veterans Administration Medical Center, Boston, Massachusetts, USA

STOCKLEY R.A.
Lung Immunobiochemical Research Laboratory, The General Hospital, Birmingham, UK

STOLK J.
Department of Pulmonology, University Hospital of Leiden, The Netherlands

STONE P.J.
Pulmonary Center and Biochemistry Department, Boston University School of Medicine, Boston, USA

TRAVIS J.
Department of Biochemistry, University of Georgia, Athens, Georgia, USA

WOO S.L.C.
Howard Hughes Medical Institute, Baylor College of Medicine, Houston, Texas, USA

Abbreviations

α-1-Achy = α-1-Antichymotrypsin
α-1-PI = α-1-Proteinase Inhibitor
A-α(1-21) = Fibrogen A-α chain
α_1-PI = α_1-Proteinase Inhibitor
α1-PI = alpha-1-Protease Inhibitor
α1AT = α1-Antitrypsin
α2M = α2-Macroglobulin
AAT = Alpha-1-Antitrypsin
ALP = Antileukoprotease
APPA = Keto Acid Inhibitor 4-amidino-
    phenylpyruvate
ARDS = Adult Respiratory Distress Syndrome
AT III = Antithrombin III
ATS = American Thoracic Society
BAL = Bronchoalveolar Lavage
BALF = Bronchoalveolar Lavage Fluid
BLPI = Bronchial Leucocyte Protease Inhibitor

Cat G = Cathepsin G
CF = Cystic Fibrosis
CFTR = Cystic Fibrosis Transmembrane
    Conductance Regulator
CGD = Chronic Granulomatous Disease
COPD = Chronic Obstructive Pulmonary
    Disease

DIC = Disseminated Intravascular Coagulation

ELF = Epithelial Lining Fluid

FEV_1 = Forced Expiratory Volume in 1 second
FMLP = Phenylmethylsulphonyl Fluoride
FVC = Forced Vital Capacity

GMCSF = Granulocyte Macrophage Colony
    Stimulating Factor

HNE = Human Neutrophil Elastase
HNE-Dt$_{1/2}$ = HNE-Half Time Dissociation
HNE-MRI$_{50}$ = HNE-Molar Ratio Inhibitor for 50%
HUSI-II = Human Seminal Acrosin Inhibitor II

IGF-1 = Insulin-like Growth Factor I
IL6 = Interleukin 6
IL8 = Interleukin 8

kb = Kilobase
kD = Kilodalton
KDa = Kilodalton

LCt$_{1/2}$ = Half Time Lung Clearance

MAC = Alveolar Macrophage
MAGP = Microfibril Associated Glycoprotein
MLI = Mean Linear Intercept
MPI = Mucus Proteinase Inhibitor
Mr = Relative Molecular Mass

NCF = Neutrophil Chemotactic Factor
NE = Neutrophil Elastase (see HNE)
NHLBI = National Heart Lung and Blood Institute

PA = Plasminogen Activator
PAI-1 = Plasminogen Activator Inhibitor 1
PDS = Thiol Reagent 4,4'-Diothiodipyridine
PI = Proteinase Inhibitor
PMN = Polymorphonuclear Neutrophils
PPE = Porcine Pancreatic Elastase
PSTI = Pancreatic Secretory Trypsin Inhibitor

rα1AT = recombinant α1-Antitrypsin
RER = Rough Endoplasmic Reticulum
ROS = Reactive Oxygen Species
rSLPI = recombinant SLPI

SLPI = Secretory Leucocyte Proteinase
    Inhibitor

TGFβ = Transforming Growth Factor β
TIMP = Tissue Inhibitor Metalloproteinase
TNF = Tumor Necrosis Factor
TNFα = Tumor Necrosis Factor α

1. Pulmonary Emphysema: What's Going On

C. GRASSI, M. LUISETTI
Institute of Phthisiology and Respiratory Diseases, IRCCS San Matteo, University of Pavia, Italy

Introduction

Pulmonary emphysema and chronic bronchitis, usually defined by the mutual term "Chronic Obstructive Pulmonary Disease" (COPD), are clinical conditions characterized by disabling airflow limitation, productive cough, and dyspnoea. Cigarette smoking is the major risk factor associated with the development of these chronic disorders, and has assumed alarming proportions during the past fifty years. A few figures convey the magnitude of the problem: in the USA, COPD was responsible in 1987 for 354,000 hospitalizations and 74,000 deaths.

In terms of anatomic distinction, we recognize two main forms of pulmonary emphysema.

Centriacinar (or centrilobular) emphysema, is the more common form, usually associated with cigarette smoking and chronic bronchitis.

The other type, panacinar (or panlobular) emphysema, is much less frequent, often developing in juveniles, relatively independent of cigarette smoking, and commonly associated with homozygous α_1-Proteinase Inhibitor (α_1PI) deficiency.

The mutual current definition of emphysema is therefore "... a condition of the lung characterized by abnormal, permanent enlargement of airspaces distal to the terminal bronchiole, accompanied by the destruction of their walls, and without obvious fibrosis."[1]

The progress of knowledge concerning the pathogenesis of pulmonary emphysema represents an exciting and paradigmatic example of biomedical research, with a distinguished past, a bright present, and a promising future.

The Past

In 1835, Laennec first proposed a theory of partial airway obstruction, due both to inflammatory narrowing of airways and to loss of elastic recoil of the pulmonary parenchyma. This amazing intuition was preserved intact over the years. For instance, in a 1952 textbook the airtrapping mechanisms proposed by Laennec are still considered, with slight modifications, the most likely pathogenetic theories.[2] Other putative causes proposed during the first half of the 20th century were a "stress" mechanism on alveolar wall due to chronic cough, the loss of pulmonary vascular bed, and undetermined nutritional and genetic factors.

In the early sixties, two observations threw light on the biochemical bases of the loss of elastic recoil in pulmonary emphysema. In 1963 two Swedish investigators, Laurell and Eriksson,[3] first reported the homozygous α_1PI deficiency associated with early development of emphysema. The following year, Gross and colleagues in the USA demonstrated that the instillation of papain, a proteolytic enzyme, into the lungs of rat yields morphologic changes similar to emphysema in humans.[4]

This field of research has developed during the past two decades. The validity of these two observations is evidenced by the high number of papers published in this biochemical field: 1,029 from 1966 to April 1990 (*source: Medline*), with a mean of 41 papers published per year. Furthermore, this considerable amount of data led to the development of the "protease-antiprotease hypothesis" of the pathogenesis of pulmonary emphysema. This hypothesis gives a satisfactory explanation of biochemical and cellular events underlying the morphologic changes in emphysema and represents the theoretical basis for the prevention and the specific treatment of emphysema in humans.

The Present

The basic concept of the protease-antiprotease hypothesis is that pulmonary emphysema occurs because of uncontrolled digestion of interstitial elastin by elastolytic proteinases. The elastase discharged by polymorphonuclear neutrophils, human neutrophil elastase (HNE), has been proved as the most likely guilty proteinase in this process. The degradation of elastin is made possible by the overcoming of a defence barrier in the lower respiratory tract, the so-called "antielastase screen".

This barrier is mainly supplied by α_1PI, a plasma glycoprotein of 52 kDa produced by hepatocytes and, to a lesser extent, by macrophages. It is a highly polymorphic protein, with about 70 biochemical variants (the PI system), easily distinguishable by isoelectric focusing.[5] The most common variant is the type M (95% of the population) and individuals homozygous for type M (PI-MM) show a mean α_1PI plasma concentration of 2.86 g/l.[6] By virtue of its relatively low molecular mass, α_1PI easily transudes from blood into the lungs, making up the largest part of the antielastase screen of the epithelial lining fluid. At this site, α_1PI binds to HNE with a high association rate constant (10^{-7} M^{-1} sec^{-1}) and inhibits the degradation effects of HNE on interstitial elastin.

Thus, the protease-antiprotease hypothesis presupposes an overcoming of such a highly efficient inhibitor. That is easily conceivable in α_1PI deficient subjects showing the defective variant Z. The homozygous individuals (PI-ZZ) have a mean α_1PI plasma concentration that is 16% that of PI-MM individuals. Consequently, α_1PI in nearly undetectable in bronchoalveolar lavage fluid (BALF) recovered from lungs of α_1PI deficient subjects.[7] According to this evidence, it is therefore conceivable that HNE, discharged by low numbers of neutrophils, is enough to induce the unrestrained degradation of interstitial elastin (Fig.1, right side). Nevertheless, the hypothesis remains to be proved in vivo. There are a considerable number of α_1PI deficient subjects, both smokers and nonsmokers, showing a preserved lung function over the years.[8] Therefore, in the pathogenesis of emphysema in α_1PI deficient individuals, defence mechanisms towards HNE other than α_1PI should also be taken into account, as well as risk factors other than HNE load.

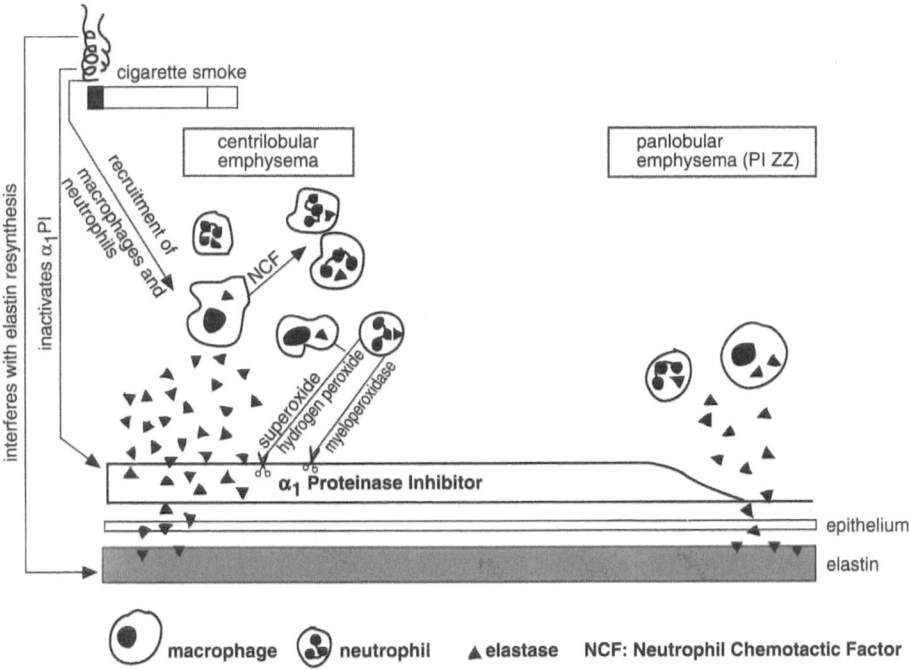

Fig. 1. Schematic representation of mechanisms involved in the pathogenesis of pulmonary emphysema according to the protease-antiprotease and oxidation hypotheses.

As previously stated, the inherited α_1PI deficiency is responsible for a minority (1 to 5 %) of cases of emphysema, most of the cases occurring in α_1PI sufficient individuals. In such individuals, the overcoming of antielastase screen in the alveolar spaces may be explained by more complex mechanisms leading to the protease-antiprotease imbalance (Fig.1, left side). In this mosaic, a key role is played by the presence of a methionine residue in the P_1 position of the active site (Met[358]) of α_1PI. A large number of studies[9] demonstrated that this residue is readily converted to methionine sulphoxide derivative under the influence of a variety of oxidant agents. As a result, a 2,000 fold decrease in the association rate constant towards HNE occurs.[10] Since cigarette smoke contains a heavy burden of oxidant agents and cigarette smoking is the major risk factor for development of emphysema, it is likely that cigarette smoke plays a crucial role in the development of emphysema in α_1PI sufficient individuals. This hypothesis is supported by the following evidence:

1. cigarette smoke, containing several oxidant species, may *in vitro* inactivate α_1PI;[11]
2. cigarette smoking is associated with increased numbers of phagocytic inflammatory cells within alveolar spaces[12] and pulmonary microvessels[13] in humans. It is well known that these cells are capable of discharging oxidant species and elastolytic enzymes;
3. cigarette smoke impairs elastin resynthesis in the lung;[14]
4. cigarette smoke may make worse the elastase-induced emphysema in animal models.[15]

According to such evidence, an "oxidation hypothesis" has been developed[16], introducing the concept of an acquired deficiency of α_1PI that makes possible the overcoming of the antielastase screen by elastase(s) discharged by inflammatory cells into the lungs. Nevertheless, a number of key questions must be answered, before a protease-antiprotease hypothesis in α_1PI sufficient individuals may be considered fully convincing. First of all, the proof that the oxidation mechanism is really acting *in vivo* is so far lacking. In fact, conflicting data began to emerge when *in vivo* investigations were carried out. Some authors described a functional decrease of α_1PI in BALF recovered from smokers with respect to nonsmokers, while other authors failed to confirm this evidence. This controversy seems far from resolved and the recent papers of Gast and colleagues[17] and Wewers and colleagues[18] are an example of these discrepancies. Next, it is possible to suppose that the concept of an imbalance of 1 protease (HNE) and 1 antiprotease (α_1PI) is too simplistic.[19]

The role of other inhibitors (α_2-macroglobulin, antileukoprotease, α_1-antichymotrypsin), and other elastases (cathepsin G, proteinase 3, macrophage elastase and bacterial elastases), so far poorly investigated *in vivo*, should be clarified. Next, apart from the oxidative inactivation, even the proteolytic inactivation of α_1PI, caused by some enzymes deriving from inflammatory cells and bacteria[16], may be a mechanism playing a role *in vivo* in the pathogenesis of emphysema.[20] The exact contribution of proteolytic inactivation to the acquired deficiency of α_1PI remains to be assessed. Finally, the importance of genetic factors other than α_1PI gene disorders is to date poorly understood, both in terms of protective gene(s) (only 1 out of 10 smokers develops disabling airflow limitation), and in terms of risk gene(s) (there is a small percentage of α_1PI sufficient COPD patients who never smoked). So far we possess only scattered information on this issue,[21,22] but it seems to be a promising and exciting field for future investigations.

Both the protease-antiprotease and the oxidation hypotheses represent the starting point for the development of specific preventive and therapeutic strategies to be combined with the traditional treatment of pulmonary emphysema (Table I).[24,26] At the present time, the only specific treatment currently performed (in USA) is augmentation therapy with α_1PI purified from human plasma and intravenously administered.[27]

This product was licensed by the US FDA in December 1987 for use in individuals with inherited deficiency of α_1PI. The ATS statement[28] suggests reserving this treatment for PI-ZZ, PI-Z null and PI-null null subjects and a few rare phenotypes associated with very low α_1PI plasma levels. It is not recomended either for deficient patients with no lung functional impairment or for emphysematous individuals with normal α_1PI phenotypes or with deficient phenotypes, but with plasma α_1PI levels above the protective threshold of 0.8 g/l. The goal of the treatment is to stop the progression of the destructive lung disease, without expecting an improvement of the lung function during the therapy. The long-term efficacy of the replacement treatment remains to be demonstrated, since a controlled clinical trial could not be carried out, mainly because the cost could not be met. A study of short-term administration of inhalatory α_1PI to deficient subjects has been recently carried out,[29] suggesting that such an approach may be an alternative to intravenous administration.

The Future

All other strategies deriving from the protease-antiprotease and the oxidation hypotheses belong to the future, near or distant. As far as α_1PI augmentation therapy is concerned, molecular biology techniques enable today the production of recombinant non-glycosylated human α_1PI ($r\alpha_1$PI).[30] This method of production allows unlimited amount of protein free from potential blood contaminants. In view of the shorter plasma half-life of $r\alpha_1$PI with respect to normal α_1PI, due to the lack of carbohydrate side-chains, the aerosol seems to be the best way of administration. A preliminary pilot study in deficient subjects showed an augmentation of antielastase

Table I. Current and future therapeutic approach to pulmonary emphysema

Pathophysiology*	Current therapy		Future therapy
	*Classic approach**	*Approach suggested by protease-antiprotease hypothesis*	
Bronchial inflammation	Corticosteroids Antibiotics		Other "antiinflammatory" drugs
Excessive airway secretions	Tracheobronchial toilet Postural drainage Mucolytic agents (Beta adrenergic agonists)		Thiolic elastase inhibitors
Bronchospasm	Methylxanthines Beta adrenergic agonists Corticosteroids Anticholinergic agents		
Loss of lung elastic recoil and prevention of further deterioration	Breathing exercise Cessation of smoking	α_1PI augmentation	α_1PI augmentation Gene therapy Elastase inhibitors other than α_1PI Antioxidant screen augmentation
Arterial hypoxemia	Supplemental low-flow oxygen		
Fluid retention and congestive heart failure	Diuretics Digitalis Supplemental low-flow oxygen		
Excercise intolerance and catabolic state	Patient education Low-flow oxygen High caloric diet		
End-stage respiratory failure	Intensive care remedies Lung transplantation		

* After Wanner and Sackner, [23] modified.

capacity of epithelial lining fluid, with no adverse reactions.[27]

Another α_1PI augmentation strategy derived directly from the oxidation hypothesis and prompted the investigators to construct rα_1PI with the oxidable Met[358] residue replaced by more resistant residues.[31,32] rα_1PI variants with Met[358] replaced with valine or leucine have been shown to be more resistant to a variety of oxidant agents and are candidates for replacement therapy in smoker α_1PI deficient subjects or in α_1PI sufficient individuals with obvious oxidative inactivation of native α_1PI in their lower respiratory tract. Immunogenic consequences of the administration of variant rα_1PI are yet unknown.

The most ambitious strategy of the α_1PI augmentation therapy is the gene replacement therapy in deficient subjects. The α_1PI gene has been inserted into mouse fibroblasts that have been then transplanted i.p. in nude mice.[33,34]

As a result, human α_1PI was detectable after 4 weeks in blood and lungs of these mice. Obviously, there are still many objections, the main being related to the concomitant expression of the endogenous mutant allele.[35]

An alternative strategy, currently not feasible *in vivo*, would be the insertion of a second mutation to correct the point mutation that codes for the Glu to Lys substitution at the residue 342 of α_1PI Z variant.[36,37] Another approach to the antielastase screen augmentation in the lower respiratory tract is to administer antielastase agents other than α_1PI.

This approach may be useful for α_1PI sufficient individuals who cannot stop smoking. There is a large variety of elastase inhibitor agents that are candidates for this approach, both of natural and synthetic origin. Among natural inhibitors, antileukoprotease (that has also been produced by recombinant DNA) shows the advantage of inhibiting elastin-bound HNE, a property that is not shared by α_1PI.[38,39]

Another powerful natural elastase inhibitor derived from medical leech, Eglin C, has unfortunately raised serious allergic problems. Synthetic inhibitors are currently receiving much attention from the investigators,[40] particularly in view of their low molecular size, that makes them suitable for displaying inhibitory activity in compartments, such as neutrophil-substrate interface, not easily accessible to higher molecular size inhibitors. Among synthetic inhibitors, the most powerful agents (such as the chloromethyl ketone irreversible inhibitors) are highly toxic and not suitable for administration in humans.

Reversible inhibitors, such as boronic acids, have instead a greater therapeutic potential, despite relatively moderate efficiency, in view of their low toxicity. Interestingly, some of these agents possess a thiolic chemical structure, predicting also a possible activity on physicochemical changes of bronchial secretions of COPD patients.[41,42]

Another promising class of antielastases is represented by modified β-lactam antibiotics.[43] At the present time, the experience on all these inhibitors other than

α_1PI is limited to *in vitro* assays and to animal models of HNE-induced emphysema.[44]

An alternative approach deriving from the protease-antiprotease and oxidation hypotheses is the reduction of elastase and oxidant burden in lungs of smokers developing emphysema. Cohen and colleagues[45,46] are currently studying a number of drugs, mainly anti-inflammatory agents, in order to evaluate their ability in reducing the release of HNE and myeloperoxidase (the latter involved in the oxidative pathway of α_1PI) from azurophilic granules of neutrophils. They found that some drugs (colchicine, auranofin, sulfinpyrazone, and phenylbutazone) have potential for therapeutic use, without affecting the other defence mechanisms of neutrophils. Antioxidant supplementation, to prevent the α_1PI oxidation, is another future strategy in the treatment of pulmonary emphysema.[47,48]

That strategy has been so far little investigated, in view of controversies on the oxidation hypothesis and on the antioxidant defensive status in smokers and COPD patients.[49,50]

Another problem is the difficulty in delivering a significant amount of active agents to the lungs. A method to improve significantly their access to the lower respiratory tract is the delivering of antioxidant agents (free or liposome-encapsulated) in aerosol form by nebulizers. Nevertheless, the efficacy of such an approach remains to be proved in animal models.

Conclusions

The main drawback to the protease-antiprotease and the oxidation hypotheses is the lack of the proof of an *in vivo* occurrence. Nevertheless, these hypotheses seem to be strong enough to suggest a brand new approach to the treatment of emphysema, one of these strategies, α_1PI i.v. administration, being currently used. The evaluation of the efficacy of antielastase treatment is likely to represent the chance of testing the validity of these hypotheses. Since an outcome study does not seem feasible because of the prohibitive cost,[44] the efforts of the investigators are now directed towards identifying at-risk smokers and blood and/or urine markers, such as elastin-derived peptides, that witness the occurrence of a degradation of lung elastin and the possibility of counteracting it by means of antielastase agents.

References

1. Snider G.L., Kleinerman J., Thurlbeck W.M., Bengali Z.H.: The definition of emphysema. Report of a National Heart, Lung, and Blood Institute, Division of Lung Diseases Workshop. Am. Rev. Respir. Dis. 1985; 132: 182-185

2. Marshall G., Perry K.M.A.: *Diseases of the Chest*. London: Butterworth & Co., 1952

3. Laurell C.B., Eriksson S.: The electrophoretic alpha1-globulin pattern of serum in α1-antitrypsin deficiency. Scand. J. Clin. Lab. Invest. 1963; 15: 132-140

4. Gross P., Babjak M.A., Tolker E., Kaschak M.: Enzymatically produced pulmonary emphysema; a preliminary report. J. Occup. Med. 1964; 6:481-484

5. Travis J., Salvesen G.: Human plasma proteinase inhibitors. Ann. Rev. Biochem. 1983; 52:655-709

6. Hutchinson D.C.S.: Epidemiology of α−1-protease inhibitor deficiency. Eur. Respir. J. 1990; 3 (suppl. 9): 29s-34s

7. Gadek J.E., Fells G.A., Zimmerman R.C., Rennard S.I., Crystal R.G.: Antielastases of the human alveolar structures. Implications for the protease-antiprotease theory of emphysema. J. Clin. Invest. 1981; 68:889-898

8. Silverman E.K., Pierce J.A., Province M.A., Pao D.C., Campbell E.J.: Variability of pulmonary function in α1-antitrypsin deficiency: clinical correlates. Ann. Intern. Med. 1989; 111:982-991

9. Janoff A.: Elastases and emphysema. Current assessment of the protease-antiprotease hypothesis. Am. Rev. Respir. Dis. 1985; 132:417-433

10. Beatty K., Bieth J., Travis J.: Kinetics of association of serine proteinases with native and oxidized α−1-proteinase inhibitor and alpha1-antichymotrypsin. J. Biol. Chem. 1980; 225:3931-3934

11. Pryor W.A., Dooley M.M., Church D.F.: Inactivation of human plasma α1-proteinase inhibitor by gas phase cigarette smoke. Biochem. Biophys. Res. Commun. 1984; 122: 676-681

12. Hunninghake G.W., Crystal R.G.: Cigarette smoking and lung destruction: accumulation of neutrophils in the lungs of cigarette smokers. Am. Rev. Respir. Dis. 1983; 128:833-838

13. MacNee W., Wiggs B., Belzberg A.S., Hogg J.C.: The effect of cigarette smoking on neutrophil kinetics in human lungs. N. Engl. J. Med. 1989; 321:924-928

14. Janoff A.: Biochemical links between cigarette smoking and pulmonary emphysema. J. Appl. Physiol. 1983;55:285-293

15. Kimmel E.C., Winsett D.W., Diamond L.: Augmentation of elastase-induced emphysema by cigarette smoking. Description of a model and a review of possible mechanisms. Am. Rev. Respir. Dis. 1985; 132:885-893

16. Travis J.: α−1-proteinase inhibitor deficiency. In Massaro D. (Ed) *Lung Cell Biology*. New York, Marcel Dekker, Inc., 1989, pp 1227-1246

17. Gast A., Dietmann-Molard A., Pelletier A., Pauli G., Bieth J.G.: The antielastase screen of the lower respiratory tract of α−1-proteinase inhibitor-sufficient patients with emphysema or pneumothorax. Am. Rev. Respir. Dis. 1990;141:880-883

18. Wewers M.D., Herzyk D.J., Gadek J.E.: Comparison of smoker and nonsmoker lavage fluid for the rate of association with neutrophil elastase. Am. J. Respir. Cell Mol. Biol. 1989;1:423-429

19. Stockley R.A.: α−1-antitrypsin and the pathogenesis of emphysema. Lung 1987; 165:61-77

20. Afford S.C., Burnett D., Campbell E.J., Cury J.D., Stockley R.A.: The assessment of α−1-proteinase inhibitor form and function in lung lavage fluid from healthy subjects. Biol. Chem. Hoppe-Seyler 1988; 369:1065-1074

21. Poller W., Meisen C., Olek K.: DNA polymorphisms of the α1-antitrypsin gene region in patients with chronic obstructive pulmonary disease. Eur. J. Clin. Invest. 1990;20:1-7

22. Gasparini P., Savoia A., Luisetti M., Peona V., Pignatti P.F.: The cystic fibrosis gene is not likely

to be involved in chronic obstructive pulmonary disease. Am. J. Respir. Cell Mol. Biol. 1990;2:297-299

23. Wanner A., Sackner M.A.: *Pulmonary Diseases. Mechanisms of altered structure and function.* Boston/Toronto, Little, Brown and Co., 1983

24. Campbell E.J.: Preventive therapy of emphysema. Lessons from the elastase model. Am. Rev. Respir. Dis. 1986; 134:435-437

25. Pierce J.A.: Antitrypsin and emphysema. Perspective and prospects. JAMA 1988;259:2890-2895

26. Pozzi E., De Rose V., Luisetti M.: Strategies in the pharmacologic prevention of lung emphysema. In Olivieri D., Bianco S. (Eds) *Airway obstruction and inflammation.* Prog. Resp. Res., Basel, Karger, 1990, vol. 24, pp 203-208

27. Hubbard R.C., Crystal R.G.: Augmentation therapy of alpha1-antitrypsin deficiency. Eur. Respir. J. 1990; 3(Suppl. 9):44s-52s

28. American Thoracic Society: Guidelines for the approach to the patient with severe hereditary α–1-antitrypsin deficiency. Am. Rev. Respir. Dis. 1989;140:1494-1497

29. Hubbard R.C., Brantly M.L., Sellers S.E., Mitchell M.E., Crystal R.G.: Anti-neutrophil elastase defenses of the lower respiratory tract in α1-antitrypsin deficiency directly augmented with an aerosol of α1-antitrypsin. Ann. Intern. Med. 1989; 111:206-212

30. Courtney M., Buchwalder A., Tessier L.H., Joye M., Benavente A., Balland A., Kohli V., Luthe R., Tolstoshev P., Lecocq J.P.: High-level production of biologically active human α–1-antitrypsin in *Escherichia Coli.* Proc. Natl. Acad. Sci. USA 1984;81:669-673

31. Travis J., Owen M., George P., Carrell R., Rosemberg S., Hallewell R.A., Barr P.J.: Isolation and properties of recombinant DNA produced variants of human α–1-proteinase inhibitor. J. Biol. Chem. 1985;260:4384-4389

32. Jallot S., Carvallo D., Tessier L.H., Roecklin D., Roitsch C., Ogushi F., Crystal R.G., Courtney M.: Altered specificities of genetically engineered α–1-antitrypsin variants. Protein Eng. 1986; 1:29-35

33. Garver R.I., Chytil A., Karlsson S., Fells G.A., Brantly M.C., Courtney M., Kantoff P.W., Nienhais A.W., Anderson W.F., Crystal R.G.: Production of glycosylated physiologically "normal" human α–1-antitrypsin by mouse fibroblasts modified by insertion of a human α–1-antitrypsin cDNA using a retroviral vector. Proc. Natl. Acad. Sci. USA 1987;84:1050-1054

34. Garver R.I., Chytil A., Courtney M., Crystal R.G.: Clonal gene therapy: transplanted mouse fibroblast clones express human α–1-antitrypsin gene *in vivo.* Science 1987;237:762-764

35. Perlemutter D.H., Pierce J.A.: The α–1-antitrypsin gene and emphysema. Am. J. Physiol. (Lung Cell Mol. Physiol.) 1989;257:L147-L162

36. Sifers R.N., Hordick C.P., Woo S.L.C.: Disruption of the 290-342 salt bridge is not responsible for the secretory defect of the PiZ α–1-antitrypsin variant. J. Biol. Chem. 1989;264:2997-3001

37. Brantly M., Courtney M., Crystal R.G.: Repair of the secretion defect in the Z form of the α–1-antitrypsin by addition of a second mutation. Science 1988;242:1700-1702

38. Bruch M., Bieth J.G: Influence of elastin on the inhibition of leucocyte elastase by α–1-proteinase inhibitor and bronchial inhibitor. Potent inhibition of elastin-bound elastase by bronchial inhibitor. Biochem. J. 1986;238:269-273

39. Kramps J.A., Te Boekhorst A.H.T., Fransen J.A.M., Ginsel L.A., Dijkman J.H.: Antileukoprotease is associated with elastin fibers in the extracellular matrix of the human lung. An immunoelectron microscopic study. Am. Rev. Respir. Dis. 1989;140:471-476

40. Powers J.C., Bengali Z.H.: Elastase inhibitors for treatment of emphysema. Approaches to synthesis and biological evaluation. Am. Rev. Respir. Dis. 1986;134:1097-1100

41. Luisetti M., Piccioni P.D., Donnini M., Peona V., Pozzi E., Grassi C.: Studies of MR 889, a new synthetic proteinase inhibitor. Biochem. Biophys. Res. Commun. 1989; 165:568-573

42. Baici A., Pelloso R., Horler D.: The kinetic mechanism of inhibition of human leukocyte elastase by MR 889, a new cyclic thiolic compound. Biochem. Pharmacol. 1990;39:919-924

43. Doherty J.B., Ashe B.M., Argenhight L.W., Barker P.L., Bormey R.J., Chandler G.O., Dahlgren M.E., Dorn Jr C.P., Finke P.E., Firestone P.A., Fletcher D., Hagmann W.K., Mumford R., O'Grady L., Maycock A.C., Pisano J.M., Shah S.K., Thompson K.R., Zimmerman M.: Cephalosporin antibiotics can be modified to inhibit human leukocyte elastase. Nature 1986; 322:192-194

44. Snider G.L., Stone P. J., Lucey E.C.: The specific treatment of emphysema. Eur. Respir. J. 1990;3(Suppl.9):23s-28s

45. Cohen A.B.: Treatment of the underlying cause of emphysema. Sem. Respir. Med. 1986;8:177-183

46. Stevens M.D., Miller E.J., Cohen A.B.: Search for drugs that may reduce the load of neutrophil azurophilic granule enzymes in the lungs of patients with emphysema. Exp. Lung Res. 1989;15:663-680

47. Heffner J.E., Repine J.E.: Pulmonary strategies of antioxidant defense. Am. Rev. Respir. Dis. 1989;140:531-554

48. Hubbard R.C., Crystal R.G.: Antiproteases and antioxidants: strategies for the pharmacologic prevention of lung destruction. Respiration 1986;50 (Suppl.1):56-73

49. Cantin A.M., North S.L., Hubbard R.C., Crystal R.G.: Normal alveolar epithelial lining fluid contains high levels of glutathione. J. Appl. Physiol. 1987; 63:152-157

50. Taylor J.C., Madison R., Kosinska D.: Is antioxidant deficiency related to chronic obstructive pulmonary disease? Am. Rev. Respir. Dis. 1986;134:285-289

2. Elastin and the Lung

J. M. DAVIDSON

Research Service, Department of Veterans Affairs Medical Center and Department of Pathology, Vanderbilt University School of Medicine, Nashville, Tennessee, USA

Elastin, a rubbery, fibrous connective tissue protein, plays several critical roles in the biomechanics of the lung and its associated vasculature. It is the principal component of the elastic fibre, which is a composite of the amorphous elastic protein and one or more microfibrillar components. Elastic fibres provide resiliency to numerous pulmonary structures including the alveolar wall, the bronchiolar interstitium, the medial layers of arterial and venous elements, lymphatics, and the pleura. The unusual hydrophobic properties of the protein, together with the presence of numerous intermolecular crosslinks, in particular desmosine and isodesmosine, produce a highly insoluble, durable network of polypeptide chains that function as a perfect elastomer in an aqueous environment.

That elastin is critical to lung function is best exemplified by conditions that produce its deficiency.

The autosomal recessive form of cutis laxa is a disease characterized by the absence or derangement of elastic fibre organization in the skin and other elastic tissue.[1] In the most severe, perinatal form of the disease, infant mortality arises from pulmonary dysfunction resembling emphysema.[2] Although the precise genetic defect(s) are unknown, reduced capacity of fibroblasts to produce elastin is characteristic of several cutis laxa probands.[3,4]

The second indication of the involvement of elastin in lung function is the characteristic loss of elastic fibres in the alveolar interstitium of humans suffering from acquired and genetically-determined pulmonary emphysema.[5] Excessive or uncontrolled action of elastolytic proteinases[6], over the course of many years, causes erosion and eventual loss of elastin with apparently little functional replacement.

Elastin Structure and Synthesis

Elastin is the most insoluble protein in the body. Its primary structure, deduced from both protein and nucleic acid sequencing, consists of 12-13 repeats of alternating hydrophobic and crosslinking domains.[7] The hydrophobic regions consist of numerous repetitive polypeptide sequences, ranging from 3-9 amino acids in length and predominantly consisting of glycine, alanine, proline, and valine residues. Although there is some potential for β-turn structure in these regions, which has led to the proposed "vibrational" form of entropic behavior,[8] other thermodynamic considerations favour the random-coil behaviour of these regions, at least when the molecules are in a relaxed state.[9] On the other hand, the crosslink sites in elastin are essentially polyalanine tracts with pairs of lysyl residues embedded in a rigid α-helical structure. Crosslinking of elastin chains occurs extracellularly when lysyl oxidase, a copper-dependent enzyme,[10] oxidatively deaminates three of four lysyl residues in two adjacent crosslink regions. Spontaneous formation of aldol condensation products with the fourth lysyl residue slowly leads to the generation of characteristic, nonreducible crosslinks, the desmosines, which can be used as a quantitative measure of insoluble elastin formation[11] (Fig.1). Elastin is synthesized and secreted from cells, fibroblasts and vascular smooth muscle cells in particular, as a soluble precursor called tropoelastin. Unlike the secretory pathway of another fibrous protein, procollagen,[12] there is no convincing evidence of processing of a more soluble precursor, but it is conceivable that a chaperonin, or molecular chaperone protein,[13] may prevent the intracellular aggregation of tropoelastin monomers during synthesis and secretion. Both tropoelastin and various soluble forms of elastin spontaneously form hydrophobic aggregates under physiologic conditions. Like collagen, tropoelastin contains hydroxyproline residues, but there is no evidence of a triple-helical conformation, and excessive ascorbate actually reduces absolute levels of elastin production[14] and accumulation.[15,16]

The gene for elastin is highly dispersed, the 3.5kb mRNA being encoded by nearly 40kb of genomic DNA.[17] Crosslink and hydrophobic exons are separate and widely spaced by intron sequences. In man, the elastin gene is located on human chromosome 7,[18] and alternative splicing of several different exons leads to limited protein polymorphism.[19]

The function of alternative forms of tropoelastin is under investigation, with some suggestion that preferential splicing patterns may mark certain developmental states;[20] however, polymorphism of tropoelastin monomers is very common,[21] and may be a mechanism whereby lateral alignment of tropoelastin monomers is retarded relative to the random organization of aggregated molecules. Upstream of the translation start site, the elastin gene contains a region with promoter activity, as measured by stimulated transcription of the chloramphenicol acetyl transferase reporter gene.[22]

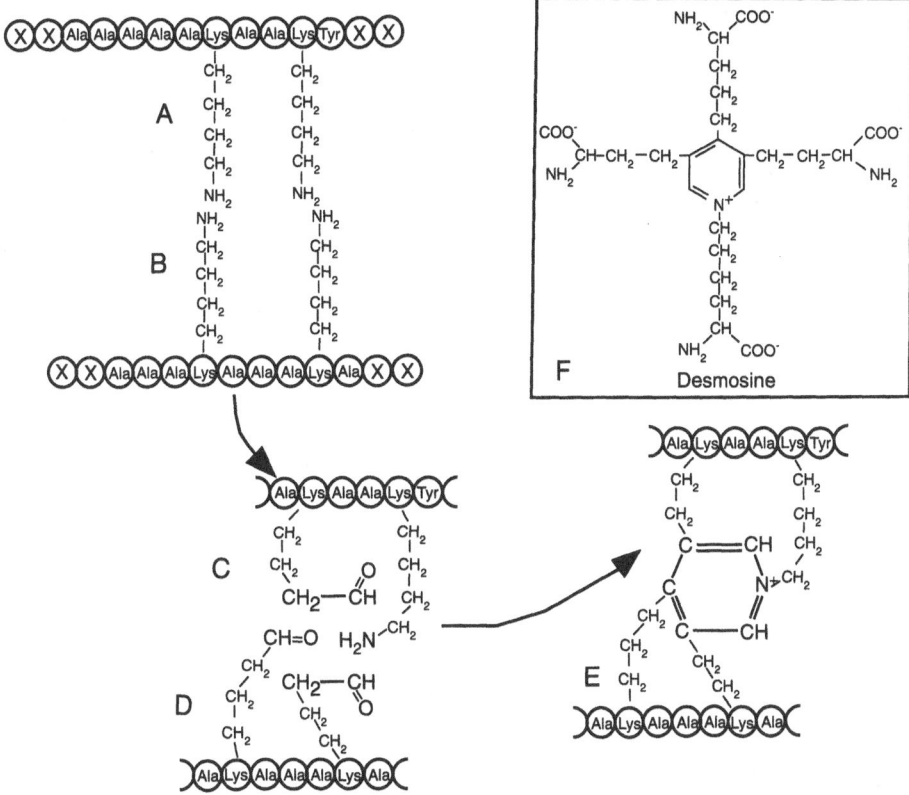

Fig. 1. Synthesis and structure of desmosine. Oxidative deamination of lysyl residues in tropoelastin leads to the spontaneous condensation of adjacent aldehydes to form the chemical structures isodesmosine and desmosine, diagnostic features of the elastin molecule. Production of this crosslinking is blocked by lathyrogens, such as β-aminopropionitrile, copper deficiency, and aldehyde blocking agents such as penicillamine. Adapted from reference 5, with permission of the publisher.

Elastic Fibre Microfibrils

The second morphologic component of the elastic fibre is an aggregate of 15 nm microfibrils.[23] During development, deposition of microfibrils precedes accumulation of visible amorphous elastin, leading to the proposed role of microfibrils in initiating the organization of elastic fibres.[24] In the upper part of the dermis, these microfibrils extend upward beyond the elastic fibre towards the epidermal basement membrane zone. A highly insoluble, cysteine-rich glycoprotein known as fibrillin has been immunologically identified in microfibrils both together with and separate from elastin.[25] This 350kDa molecule has a high propensity to aggregate, making determination of its complete primary structure difficult.[26] Other glycoprotein components of microfibrils have been partially characterized, some being immunologically related to fibrillin, while others appear to have distinct biochemical properties. Prominent in this category is a 31kDa microfibril-associated glycoprotein (MAGP).[27] Little is known about the biosynthesis of fibrillin and other microfibrillar components, except that they are produced by a variety of fibroblasts, and the absence of fibrillin appears to be common in fibroblasts derived from patients with the Marfan syndrome.[28]

Elastin Binding Proteins

Cellular interaction with elastin occurs during elastogenesis and elastolysis. Elastin deposition, at least in cell culture, appears to be dependent on the function of a 67 kDa elastin-binding molecule associated with fibroblast plasma membranes.[29,30]

The binding protein is thought to be noncovalently linked to two other integral membrane proteins. Although the recognition site of this peripheral membrane protein appears to be a distinctive hydrophobic sequence in elastin (VGVAPG), the same binding protein apparently reacts with laminin through a related sequence LGTIPG.[31] Possible functions of the elastin/laminin binding protein include chemotactic response, scavenging of degradation products, organization of elastic fibres in the matrix, elastin secretion, and cell attachment. The VGVAPG epitope is also implicated in the trafficking of malignant cells to the lung, since the lung-specific line of Lewis lung carcinoma[32] and B16-F10 melanoma[33] show binding to insoluble elastin and chemotaxis to elastin fragments or the elastin hexapeptide. Binding in this system has been shown to occur through a very similar laminin/elastin-binding protein (Mr=59,000 kDa).

This molecular complex may also be present on monocytes, which, together with ligament fibroblasts, are specifically attracted to elastin peptides.[34,35] Neutrophils, whose role is central in current models of emphysema contain a similar 67kDa binding protein that also recognizes type IV collagen and its 7S domain.[36] Increased reactivity of mononuclear cells and polymorphonuclear granulocytes in the presence of elastin fragments may be a critical component in production of cytokines during the course of lung inflammation. An unusual feature of this binding protein is its lectin properties. It appears to be related to another galactose-binding lectin (galaptin, 14kDa lectin) described by others.[37] Binding of galactosyl ligands to the lectin portion of the molecule causes loss of elastin/laminin binding and brings about disorganization of elastin fibres in cell culture. The role of this molecule is still controversial.[38] Another 120 kDa protein, termed elastonectin,[39] has been implicated in attachment of elastic fibre fragments to cell surfaces. Elastonectin activity is reported to be induced by the presence of elastin peptides.[40] Elastin itself is not known to be a good substrate for cell attachment, probably because of its highly hydrophobic character. Despite this fact, cells such as smooth muscle are intimately entwined in an elastic fibre network and elastic fibres can be visualized very near the plasma membrane.[41] Thus other pericellular molecules may facilitate the association of cells with surrounding elastic fibres.

Elastin in Lung Development

Elastic fibres and elastin begin to accumulate in the pulmonary interstitium relatively late in lung development. In the sheep, for example, elastin mRNA levels and rates of elastin synthesis by lung explants begin to rise rapidly in the third trimester, peaking about the time of gestation[42] (Fig. 2). This is coincident with the final stage of alveolarization of the neonatal lung. By comparison, the rat and rabbit lung go through the stage of alveolarization postnatally, and this is reflected in the delayed onset of maximal elastin accumulation in the newborn lung.[43] As juvenile development proceeds, elastin accumulation falls to low levels, and there is little if any data showing neosynthesis of elastin in the normal, adult lung. Indeed, turnover studies have suggested biological half-lives of between 6 and 60 years for pulmonary elastic tissue.[44] However, elastic tissue of the lung can undergo resynthesis after destructive injury.

Pathogenesis of Emphysema

Extensive study of pulmonary emphysema from both morphologic and physiologic perspectives make it clear that loss of elastic recoil is the hallmark of the disease.[45] Experimental models in animals have shown that destruction of elastic fibres by a variety of means produces a pathology resembling pulmonary emphysema in humans.[5]

Taken together with the high probability of α-1 antiprotease-deficient patients developing the disease,[46] most investigators accept that the primary lesion of emphysema is the irreversible destruction of elastic fibres among other matrix components within the lung interstitium, leading to loss of a major functional component of the extracellular matrix.

Regulation and Repair of Elastic Tissue

A key problem in emphysema is understanding why the lung is unable to adequately repair the damage to elastic fibres under certain circumstances.[47] It may be that the rate of proteolysis simply exceeds the rate of synthesis of elastin, leading to net loss, over time, of functional elastin. However, studies from experimental animals and man show that quantitative return of elastin content occurs in both the authentic disease and its models.[48] This would suggest that the lung senses and can

18

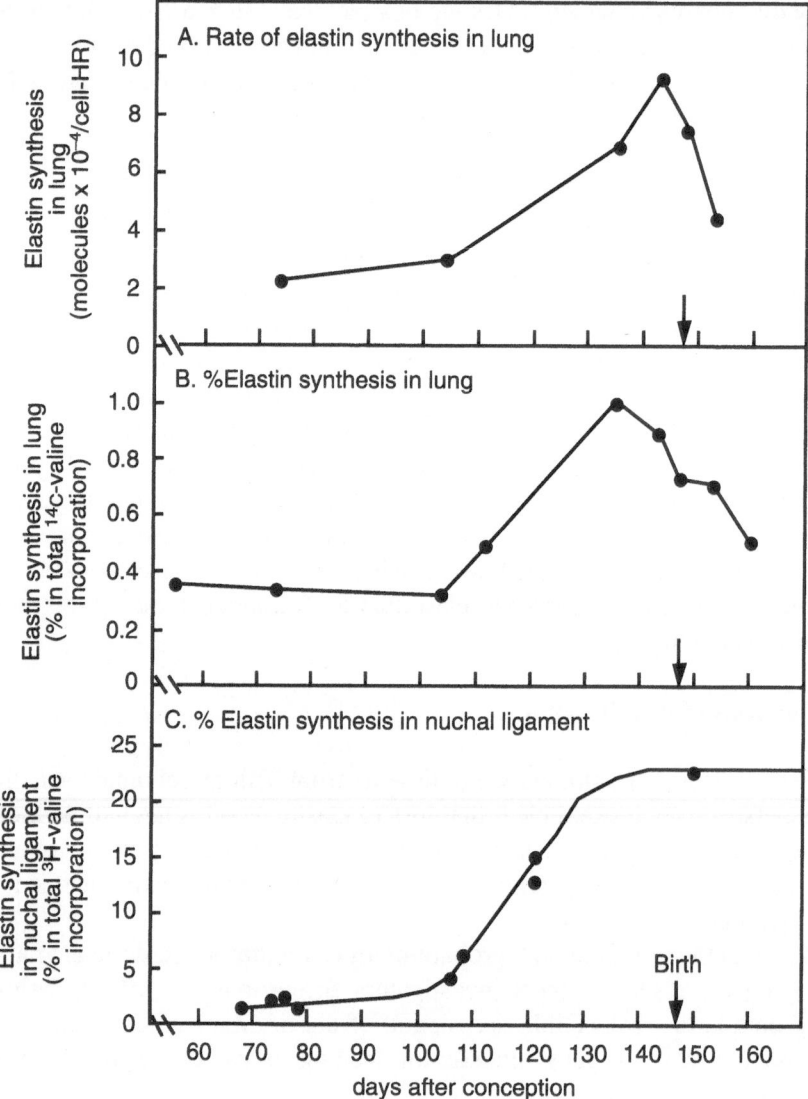

Fig. 2. Developmentally regulated expression of elastin in lung and nuchal ligament of the fetal sheep. Absolute (A) and relative (B,C) rates of elastin synthesis were determined in organ culture of explanted lung and nuchal ligament tissue. Both absolute and relative rates declined before term in this species. Adapted from reference 42 with permission.

mount a biosynthetic response to loss of elastin, but that proper arrangement of elastic fibres and other components of the damaged alveolar wall is not achievable once the alveolar template is obliterated. During development, lung tissue goes through a complex morphogenetic process,[49] and it is proposed that the healing of the emphysematous lung is a form of scarring with loss of function.

That injury can invoke neosynthesis of elastin is evidenced in at least two other experimental systems: cutaneous wounds,[50,51] and experimental pulmonary hypertension.[52,53] In both cases, damaged tissue rapidly responds by increasing levels of elastin mRNA,[54] presumably leading to the accumulation of protein. The specific signal for increased elastin production in injured tissue is uncertain and may depend upon the model or disease studied. Transforming growth factor-β is one likely mediator of elastin accumulation at sites of injury.[51] This is supported by three lines of evidence. First, TGF-β added directly to cultures of fibroblasts or smooth muscle cells strongly stimulates elastin[55] (Fig. 3), together with other matrix components

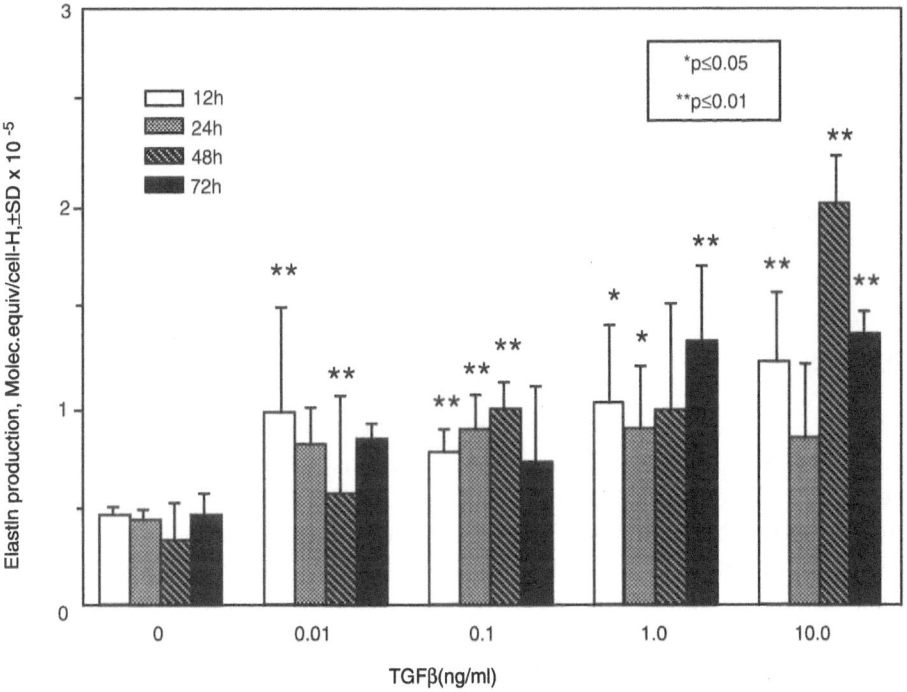

Fig. 3. Transforming growth factor-beta (TGFβ) induces a large increase in SMC elastin production. Confluent SMC were changed to medium containing 10% newborn calf serum and the indicated concentrations of recombinant human TGFβ (provided by Rik Derynck, Genentech, Inc.). Media were collected after 12-72 h and analyzed for elastin production. Significance was tested by the Student's t-test for independent means on triplicate dishes of cells and duplicate assays. Levels of TGFβ as low as 10-100 pg/ml elicited significant increases in elastin production under these conditions. Cell replication was low in confluent cultures of porcine SMC, and cell number in sister cultures was not significantly altered in any of these conditions. Over the dose range tested, strong induction could be seen after as little as 12h exposure to the cytokine, and SMC were stimulated up to 4-fold over control levels by doses of 10 ng/ml (0.4 pM). Like many of the other responses of cells to TGFβ, subpicomolar amounts produced significant biological changes. Adapted from reference 55 with permission of the publisher.

such as collagen and proteoglycan.[56] Second, TGF-β or its mRNA is present at sites of tissue repair such as cutaneous wounds,[50] the epithelial lining fluid of injured lungs[57], lung lymph drainage[58] and the thickening medial wall of hypertensive pulmonary artery.[59] Third, addition of exogenous TGF-β to tissues results in increased local expression of elastin[50] and interstitial collagen.[56] Thus, it is likely that the increased elastin deposited after various forms of lung injury is mediated, at least in part, by expression of TGF-β by associated inflammatory cells. In addition, insulin-like growth factor I (IGF-1; somatomedin C) has been implicated in stimulated synthesis of elastin in cultured cells,[60,61] and may be particularly important in developmentally-regulated elastin production.[62] Production of this cytokine by resident macrophages is also characteristic of chronic inflammation.[63] It is probable that no single factor determines elastin synthetic levels. Elastin production is also under complex regulation by other endogenous and exogenous influences such as glucocorticoids, the mitogenic cytokines, basic fibroblast growth factor, transforming growth factor-α, and the methylation state of the elastin gene.[14] Effective elastin deposition may also require the coordinate regulation of microfibrillar protein production, elastin binding proteins, and lysyl oxidase. Interestingly, the recent cloning of this rat lysyl oxidase has revealed a startling homology between a segment of the 3 untranslated regions of elastin and lysyl oxidase mRNA,[64] suggesting possible common regulatory elements.

Physiologic Monitoring of Elastin Metabolism

Are there effective means of monitoring the status of elastin metabolism during the course of emphysema? In animal models, the instillation of a bolus of elastase into the lung causes rapid, acute destruction of alveolar structures, extravasation of inflammatory cells, and loss of lung elastin content. Concomitantly, elastin degradation products, measured as desmosines, rise sharply and transiently in urine.[65] Similarly, patients with α1-antiprotease deficiency show elevated levels of excretion of urinary desmosine.[66] We have recently adapted an ELISA assay to the measurement of total elastin peptides in biological fluids.[67,68] Using antibodies directed against chemically-solubilized elastin, this assay has readily detected elastin peptide fragments in plasma lung lymph drainage and urine of humans and several large animal species. Our recent studies have shown that both plasma and urinary levels of elastin peptides were highly elevated in patients with chronic obstructive pulmonary disease. Since our antibody did not depend on the presence of cross-links, it has been presumed to detect the turnover of both pre-existing and newly-synthesized elastin. Coupled with a specific and sensitive assay for desmosine, we believe it will be possible to demonstrate whether old or new sources of elastin are prevalent in the urine of emphysema patients. More importantly, the ELISA assay, which appears to be more rapid, specific and reliable than desmosine immunoassays, can be used as a monitor for the efficacy of protective agents in the treatment or amelioration of various forms of pulmonary emphysema. A preliminary study of patients receiving α1-antiprotease therapy (Prolastin), conducted by Gordon Bernard at the Vanderbilt University Centre for Lung Research, has shown a 30% reduction in urinary elastin peptide excretion during 9 weeks of treatment.

Lung fluids contain numerous mediators that may be linked to pathophysiologic processes. Bronchoalveolar lavage fluid, or epithelial lining fluid, has been shown to contain not only structural proteins and their degradation products[69], but active mediators of lung cell behaviour, such as TGF-β and IGF-I-like activity.[57,63] In sheep models, used extensively for studies of acute lung injury and pulmonary hypertension, we and our colleagues have detected significant changes in mitogenic and elastogenic activities present in the lymph drainage of the sheep lung interstitium.[58,59,70] A substantial fraction of the elastogenic activity in lymph is attributable to TGF-β.

Perspectives

A number of aspects of elastin in the lung remain uncertain. The specific cellular sources of elastin during ontogeny and pathology are not defined. *In vitro*, elastin can be produced by vascular smooth muscle cells, fibroblasts, mesothelial cells, and even vascular endothelium.[71,72] Based on its location within the alveolar interstitium, the interstitial fibroblast is a critical target cell in understanding emphysema. The control of elastin synthesis is only partly understood. If emphysema indeed results from the imbalance between synthetic and degradative processes, it is conceivable that alveoli might be protected by agents which stimulate local elaboration of elastin. Indeed, it may even be possible to introduce either extra copies of the elastin gene or genes whose products control the expression of elastin into emphysematous lungs as a means of offsetting the excessive destruction of this critical component. Finally, we need to develop a better understanding of the processes of tissue repair within the lung interstitium. Why do some inflammatory processes in the alveolus completely resolve while others progress to loss of functional air exchange units? Are there agents which can stimulate the recapitulation of development within damaged lungs to form alveoli *de novo*? Are there structural differences in elastin in emphysematous lungs that contribute to progressive loss of the protein?

We believe an understanding of elastin biology is central to the analysis and treatment of pulmonary emphysema as well as several other lung diseases. The peculiar chemistry and ontogeny of this fibrous protein present challenges for the biochemist and cell biologist.

Supported in part by the Department of Veterans Affairs and NIH grants GM37387 and AG06528

References

1. Uitto J., Olsen D.R., Fazio M.J., Rosenbloom J.: Molecular pathology of elastin. In: Tamburro A.M., Davidson J.M. (Eds.) *Elastin: Chemical and Biological Aspects.* Galatina, Congedo Editore, 1990; 305-330

2. Ledoux-Corbusier M.: Cutis laxa: congenital form with pulmonary emphysema. An ultrastructural study. J. Cutan. Pathol. 1983; 10:340-349

3. Sephel G.C., Byers P.H., Holbrook K.A., Davidson J.M.: Heterogeneity of elastin expression in cutis laxa fibroblast strains. J. Invest. Dermatol. 1989; 93:147-153

4. Olsen D.A., Fazio M.J., Shamban A.T., Rosenbloom J., Uitto J.: Cutis laxa-reduced elastin gene expression in skin fibroblast cultures as determined by hybridization with a homologous cDNA and an exon 1-specific oligonucleotide. J. Biol. Chem. 1988;263: 6465-6468

5. Soskel N.T., Sandberg L.B.: Pulmonary emphysema. From animal models to human diseases. In: Uitto J., Perejda A., (Eds.) *Connective Tissue Disease,* New York, Marcel Dekker, 1986; 423-453

6. Janoff A.: Elastase in tissue injury. Annu. Rev. Med. 1985; 36:207-216

7. Bashir M., Indih Z., Yeh H., Abrams W., Ornstein-Goldstein N., Rosenbloom J.C., Fazio M., Uitto J., Mecham R., Rosenbloom J.: Elastin gene structure and mRNA alternative splicing. In: Tamburro A.M., Davidson J.M. (Eds.) *Elastin: Chemical and Biological Aspects,* Galatina, Congedo Editore 1990; 45-70

8. Urry D.W.: Entropic elastic processes in protein mechanisms. I. Elastic structure due to an inverse temperature transition and elasticity due to internal chain dynamics. J. Protein. Chem. 1988; 7:1-34

9. Gosline J.M.: Hydrophobic interaction and a model for the elasticity of elastin. Biopolymers 1978;17: 677-695

10. Kagan H.M.: Characterization and regulation of lysyl oxidase. In: Mecham R.P., (Ed.) *Regulation of Matrix Accumulation* New York, Academic Press., 1986;322-399

11. Starcher B.C.: Elastin and the lung. Thorax 1986; 41:577-585

12. Davidson J.M., Berg R.A.: Posttransational events in collagen biosynthesis. In: Hand A., Oliver C. (Eds.) *Meth. Cell Biol.* New York, Academic Press 1981; 119-136

13. Ellis R.J., van der Vies S.M., Hemmingsen S.M.: The molecular chaperone concept. Biochem. Soc. Symp. 1989; 55:145-153

14. Davidson J.M., Giro M.G., Sutcliffe M., Zoia O., Quaglino Jr. D., Liu J.M., Perkett E., Meyrick B., Broadley K.M., Russel S., Sephel G.C.: Regulation of elastin synthesis. In: Tamburro A.M., Davidson J.M. (Eds.) *Elastin: Chemical and Biological Aspects,* Galatina, Congedo Editore 1990; 393-405

15. De Clerck Y.A., Jones P.A.: The effect of ascorbic acid on the nature an α production of collagen and elastin by rat smooth muscle cells. Biochem J. 1986:217-225

16. Dunn D.M., Franzblau C.: Effects of ascorbate on insoluble elastin accumulation and crosslink formation in rabbit pulmonary artery smooth muscle cell cultures. Biochemistry 1983;21:4195-4202

17. Indik Z., Yeh H., Ornstein-Goldstein N., Sheppard P., Anderson N., Rosenbloom J.C., Peltoman L., Rosenbloom J.: Alternative splicing of human elastin mRNA indicated by sequence analysis of cloned genomic and complementary DNA. Proc. Natl. Acad. Sci. USA 1987; 84:5680-5684

18. Fazio M.J., Mattei M.G., Chu M.L., Black D., Solomon E., Davidson J.M., Uitto J.: Human elastin gene: Chromosomal mapping to locus 7q11.2. Am. J. Hum. Genet. 1991; 48: 696-703

19. Raju K., Anwar R.A.: Primary structures of bovine elastin a,b, and c deduced from the sequences of cDNA clones. J. Biol. Chem. 1987; 262:5755-5762

20. Pollock J., Baule V.J., Rich C.B., Ginsburg C.D., Curtis S., Foster J.A.: Chick tropoelastin isoforms. From the gene to the extracellular matrix. J. Biol. Chem. 1990; 265:3697-3702

21. Wreen D.S., Parks W.C., Whitehouse L.A., Crouch E.C., Kinich U., Rosenbloom J., Mecham R.P.: Identification of multiple tropoelastins secreted by bovine cells. J. Biol. Chem. 1987; 262:2244-2249

22. Fazio M.J., Kahari V.M., Bashir M.M., Saitla B., Rosenbloom J., Uitto J.: Regulation of elastin gene expression: evidence for functional promoter activity in the 5-flanking region of the human gene. J. Invest. Dermatol. 1990; 94:181-196

23. Cleary E.G.: The microfibrillar component of the elastic fibers. Morphology and Biochemistry. In: Uitto J., Perejda A. (Eds.) *Connective Tissue Disease* . New York, Marcel Dekker 1986;55-81

24. Ross R., Bornstein P.: Elastic fibers in the body. Sci. Am. 1971;224:44-52

25. Sakai L.Y., Keene D.R., Engvall E.: Fibrillin, a new 350kD glycoprotein, is a component of extracellular microfibrils. J. Cell. Biol. 1986;103:2499-2509

26. Maddox B.K., Sakai L.Y., Keene D.R., Glanville R.W.: Connective tissue microfibrils. Isolation and characterization of three large pepsin-resistant domains of fibrillin. J. Biol. Chem.1989; 264: 21381-21385

27. Gibson M.A., Kumaratilake J.S., Cleary E.G.: The protein components of the 12-nanometer microfibrils of elastic and nonelastic tissue. J. Biol. Chem. 1989;264:4590-4598

28. Hollister D.W., Godfrey M., Sakai L.Y., Pyeritz R.E.: Immunohistologic abnormalities of the microfibrillar-fiber system in the Marfan syndrome. N. Engl. J. Med. 1990; 32:152-159

29. Hinek A., Wrenn D.S., Mecham R.P., Barondes S.H.: The elastin receptor, a galactoside-binding protein. Science 1988; 239:1539-1541

30. Mecham R.P., Hinek A., Entwistle R., Wrenn D.S., Griffin G.L., Senior R.M.: Elastin binds to a multifunctional 97-Kilodalton peripheral membrane protein. Biochemistry 1989; 28: 3716-3722

31. Mecham R.P., Hinek A., Griffin G.L., Senior R.M., Liotta L.A.: The elastin receptor shows structural and functional similarities to the 67-kDa tumor cell laminin receptor. J. Biol. Chem. 1989; 264: 16652-16657

32. Blood C.H., Sasse J., Brodt P., Zetter B.R.: Identification of a tumor cell receptor for VGVAPG, an elastin-derived chemotactic peptide. J. Cell. Biol. 1988;107:1987-1993

33. Netland P.A., Zetter B.R.: Melanoma cell adhesion to defined extracellular matrix components. Biochem. Biophys. Res. Commun. 1986; 139:515-522

34. Hunninghake G.W., Davidson J.M., Rennard S.I., Szapiel S., Gadek J., Crystal R.G.: Mechanisms of pulmonary emphysema. Attraction of macrophage precursors to sites of disease activity by elastin fragments. Science, 1981; 212:925-927

35. Tobias J.W., Bern M.M., Netland P.A., Zetter B.R.: Monocyte adhesion to subendothelial components. Blood 1987; 69:1265-1268

36. Senior R.M., Hinek A., Griffin G.L., Pipoly D.J., Crouch E.C., Mecham R.P.: Neutrophils show chemotaxis to type IV collagen and its 7S domain and contain a 67 kD type IV collagen binding protein with lectin properties. Am. J. Respir. Cell Mol. Biol. 1989; 1:479-487

37. Cerra R.F., Haywood Reed P.L., Barondes S.H.: Endogenous mammalian lectin localized extracellularly in lung elastic fibers. J. Cell. Biol. 1984; 98:1580-1589

38. Powell J.T.: Evidence against lung galaptin being important to the synthesis or organization of the elastic fibril. Biochem. J. 1988; 252:447-452

39. Hornebeck W., Tixier J.M., Robert L.: Inducible adhesion of mesenchymal cells to elastic fibers. Proc. Natl. Acad. Sci. USA, 1986; 83:5517-5520

40. Robert L. Jacob M.P., Fulop T., Timar J., Hornebeck W.: Elastonectin in the elastin receptor. Pathol. Biol. (Paris) 1989; 37:736-741

41. Clark J.M., Glagov S.: Transmural organization of the arterial media. The lamellar unit revisited. Arteriosclerosis 1985; 5:19-34

42. Shibahara S., Davidson J.M., Smith K., Crystal R.O.: Modulation of elastin synthesis and mRNA

activity in developing sheep lung. Biochemistry 1981; 20:6577-6584

43. Myers B., Dubick M., Last J.A., Rucker R.B.: Elastin synthesis during perinatal lung development in the rat. Biochim. Biophys. Acta 1983; 761:17-22

44. Lefevre M., Rucker R.B.: Aorta elastin turnover in normal and hypercholesterolemic Japanese quail. Biochim. Biophys. Acta 1980; 630:519-529

45. Christie R.: The elastic properties of the emphysematous lung and their clinical significance. J. Clin. Invest. 1934; 13:295-297

46. Laurell C.N., Eriksson S.: The electrophoretic α1-globulin pattern of serum in α1-antitrypsin deficiency. Scand. J. Clin. Lab. Invest. 1963; 15:132-140

47. Kuhn C., Senior R.M.: The role of elastases in the development of emphysema. Lung. 1978; 155:185-197

48. Pierce J.A., Hocott J.B., Ebert R.V.: The collagen and elastin content of the lung in emphysema. Ann. Intern. Med. 1961; 55:210-222

49. Meyrick B., Reid L.M.:In: Hodson W.A. (Ed.) *Development of the Lung.* New York, Marcel Dekker, 1977; 135-214

50. Quaglino Jr D., Nanney L.B., Kennedy R., Davidson J.M.: Localized effects of transforming growth factor-β on extracellular matrix gene expression during wound healing. I. Excisional wound model. Lab. Invest. 1990; 63: 307-319

51. Davidson J.M., Giro M.G.: Repair of elastic tissue. In: Cohen I.K., Digelmann R., Lindblad W. (Eds.) *Wound Repair*, 3rd ed. New York, Saunders, 1991

52. Perkett E.A., Brigham L.K., Meyrick B.: Continuous air embolization into sheep causes sustained pulmonary hypertension and increased vascular reactivity. Am. J. Path. 1988;123:444-454

53. Mecham R.P., Withehouse L.A., Wrenn D.S., Parks W.C., Griffin G.L., Senior R.M., Crouch E.C., Stenmark K.R., Voelkel N.F.: Smooth muscle-mediated connective tissue remodeling in pulmonary hypertension. Science 1987; 237:423-426

54. Prosser I.W., Stenmark K.R., Suthar M., Crouch E.C., Mecham R.P., Parks W.C.: Regional heterogeneity of elastin and collagen gene expression in intralobar arteries in response to hypoxic pulmonary hypertension as demonstrated by in situ hybridization. Am. J. Path. 1989;135:1073-1087

55. Liu J.M., Davidson J.M.: The elastogenic effect of recombinant transforming growth factor-β on porcine aortic smooth muscle cells. Biochem. Biophys. Res. Commun. 1988; 154:895-901

56. Roberts A.B., Flanders K.C., Kondaiah P., Thompson N.I., Van Obberghen-Schilling E., Wakefield L., Rossi P., de Crombrugghe B., Heine U., Sporn M.B.: Transforming growth factor-β, biochemistry and roles in embryogenesis, tissue repair and remodeling, and carcinogenesis. Rec. Prog. Horm. Res. 1988;44:157-197

57. Yamauchi K., Martinet Y., Basset P., Fells G.S., Crystal R.G.: High levels of transforming growth factor-β are present in the epithelial lining fluid of the normal human respiratory tract. Am. Rev. Resp. Dis. 1988;137:1360-1363

58. Perkett E.A., Lyons R.M., Moses H.L., Brigham K.L., Meyrick B.: Transforming growth factor-β activity in sheep lung lymph during development of pulmonary hypertension. J. Clin. Invest., 1990, 86, 1459-1464

59. Perkett E.A.: *Personal communication*

60. Foster J., Rich C.B., Florini J.R.: Insulin-like growth factor, somatomedin C, induces the synthesis of elastin in aortic tissue. Coll. Rel. Res. 1987; 7:161-169

61. Badesch D.B., Lee P.D., Parks W.C., Stenmark K.R.: Insulin-like growth factor I stimulates elastin synthesis by bovine pulmonary arterial smooth muscle cells. Biochem Biophys. Res. Commun. 1989;160:382-397

62. Foster J.A., Miller M.L., Benedict M.R., Richmann R.A., Rich C.B.: Evidence for insulin-like

growth factor regulation of chick aortic elastogenesis. Matrix 1989; 9:328-335

63. Rom W.N., Basset P., Fells G.A., Nukiwa T., Trapnell B.C., Crystal R.G.: Alveolar macrophages release an insulin-like growth factor-I-type molecule. J. Clin. Invest. 1988; 82:1685-1693

64. Trackman P.C., Pratt A.M., Wolanski A., Tang S.S., Offner G.D., Troxler R.F., Kagan H.M.: Cloning of rat aorta lysyl oxidase cDNA: Complete codons and predicted amino acid sequence. Biochemistry 1990; 29:4863-4870

65. King G.S., Starcher B.C., Kuhn C.T.: The measurement of elastin turnover by the radioimmunoassay of urinary desmosine excretion. Bull. Eur. Physiopathol. Respir. 1980;16 (suppl):61-64

66. Pelham F., Wewers M., Crystal R., Buist A.S., Janoff A.: Urinary excretion of desmosine (elastin cross-links) in subjects with PiZZ α1-antitrypsin deficiency, a phenotype associated with hereditary disposition to pulmonary emphysema. Am. Rev. Respir. Dis. 1985;132:821-823

67. Schriver E.E., Bernard G.R., Swindell B.B., Sutcliffe M.S., Davidson J.M.: Elastin fragment levels in human plasma, urine, and bronchoalveolar lavage fluid. (BALF) Chest 1989;96:1535

68. Perkett E.A., Davidson J.M., Curtis-Atchley P., Sutcliffe M.C., Brigham K.L., Meyrick B.: Lung lymph elastin concentration is increased in sheep with pulmonary hypertension secondary to air embolization. Am. Rev. Respir. Dis. 1989;139:A174

69. Gadek J.W., Hunninghake G.W., Fells G.A., Zimmerman R.L., Keogh B.A., Crystal R.G.: Evaluation of the protease-antiprotease theory of human destructive lung disease. Bull. Eur. Physiopath. Resp. 1980; 16(Suppl):27-40

70. Davidson J.M., Zoia O., Nair R., Meyrick B., Perkett E.A.: *unpublished studies*

71. Davidson J.M., Giro M.G.: Control of elastin synthesis: Molecular and cellular aspects. In: Mecham R.P. (Ed.) *Regulation of Matrix Accumulation.* New York, Academic Press, 1986;177-216

72. Davidson J.M.: Biochemistry and turnover of lung interstitium. Eur. Resp. J. 1990; 3: 1048-1068

3. An Introduction to the Endopeptidases

A. J. BARRETT

Department of Biochemistry, Strangeways Research Laboratory, Cambridge, UK

Introduction

A very large number of proteolytic enzymes exist in the human body, and it is no easy matter to work out how they are involved in specific physiological and pathological processes. A concept that has proved helpful in this is that of splitting the enzymes into groups on the basis of (a) the type of reaction that they catalyse and (b) the chemical mechanism of catalysis that they use.

The Classification of Peptidases

All of the enzymes that hydrolyse peptide bonds are now known as *peptidases*, and these are divided into the *exopeptidases* and the *endopeptidases*. When peptidases act to break down a molecule composed of a chain of amino acids (which may be called an oligopeptide, a polypeptide or a protein, according to its length), the exopeptidases cleave bonds only near the ends of the chain, to release blocks of one, two or three amino acids, whereas the endopeptidases can break bonds in the inner regions of the chain. The term *proteinase* means exactly the same thing as *endopeptidase*, and the two can be used interchangeably (Fig. 1).[1]

When we wish to think about the tissue damage that occurs in pulmonary emphysema, we are concerned with the breakdown of the structural proteins of the lung, notably elastin and collagen, and it is *endopeptidases* that are responsible for starting the process of breaking down proteins, and control its rate. *Exopeptidases* act at a later stage to complete the process, and are not rate-limiting.

The endopeptidases are further divided into types according to the chemical

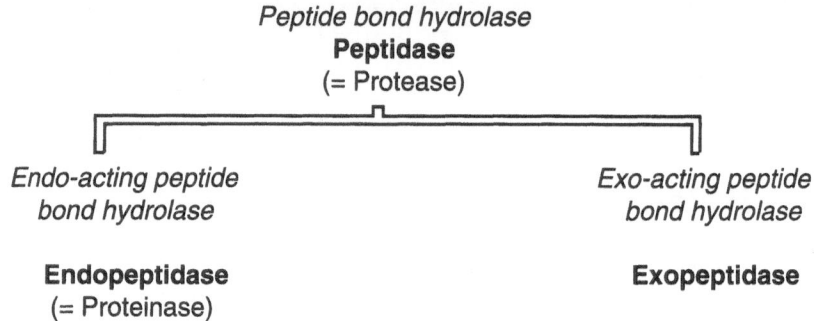

Fig. 1. The term 'peptidase', and its major divisions. For each type of activity, the systematic description is given in italics, the recommended name in bold type, and the common synonym in normal type.[1]

mechanism by which they hydrolyse peptide bonds, and these are named according to the chemical moiety directly involved in catalysis.[2] Thus, there are the *serine-type*, *cysteine-type*, *aspartic-type* and *metallo*-endopeptidases. (The word '-type' is included when necessary to make a clear distinction between catalytic type and specificity for a particular amino acyl bond). The classification is helpful, because the members of each group have many of the same properties, so we do not have to learn these properties anew for each enzyme, and can concentrate on the significant differences.

A Word about Control Mechanisms

The endopeptidases are potentially very damaging to the body, and certainly there are a number of endopeptidases in the body that are potentially quite capable of degrading the structural proteins of the lung, given conditions favourable for their activity. Not surprisingly, strict control systems exist for the enzymes, and I would suggest that it is an understanding of the way in which the enzymes are controlled, and sometimes released from control, that may be expected to lead to an eventual understanding of tissue damage in the lung.

The recognition of the catalytic types of endopeptidases is especially useful in connection with the understanding of pathophysiological situations because control mechanisms tend to be similar for endopeptidases of any one catalytic type (Table I).

Endopeptidases are generally synthesised as inactive precursors that require activation by specific, limited proteolysis. If this were not so, they would tend to digest the biosynthetic organelles of the cells in which they were formed. Once activated, they are either restricted to activity at an acidic pH value that is produced only in restricted locations in the body (most of the aspartic and cysteine

Table I. A summary of the mechanisms of regulation of extracellular activity for the various types of endopeptidase

Serine	Most need proteolytic activation; excess inhibitors are present: α_1PI, α_2M
Cysteine	Need proteolytic activation? Lack thiol activator; pH generally too high; inhibitors: cystatin C, Kininogens, α_2M
Aspartic	Most need proteolytic activation; pH generally too high
Metallo	Need proteolytic activation;excess inhibitors are present: TIMP's, α_2M

endopeptidases), or they are sensitive to inhibitors that tend to be present in large molar excess in most extracellular locations (most of the serine and metallo-endopeptidases), or both forms of control may apply. In the healthy state, the controls on activation and activity of the endopeptidases are so rigorous that for some it is quite difficult to see how they ever get an opportunity to function. In inflammation, the balance between activity and inhibition of the endopeptidases shifts decisively towards activity, however.

To complete the introduction of the agents that contribute to the balance between the endopeptidases and their inhibitors, it is necessary to mention one very exceptional inhibitor, α_2-macroglobulin. Almost all of the protein inhibitors of proteolytic enzymes in the body, and indeed in nature, affect endopeptidases of only one catalytic type, but this is not the case for α_2-macroglobulin. This is a plasma protein of high molecular weight which differs from almost all the other inhibitors produced in the body by the fact that it is capable of blocking the activity of endopeptidases of all four catalytic types.[3] The large molecular size of α_2-macroglobulin normally prevents it from escaping from the circulation into the tissues, but the increased vascular permeability at sites of inflammation does lead to the appearance of the protein in the tissue fluids.

With the importance of control mechanisms in mind, let us return to the consideration of what endopeptidases are to be found in the body.

Serine Endopeptidases

The serine endopeptidases, depending on active site serine and histidine residues for their catalytic activity, are active at neutral to alkaline pH values. They are abundant in the body as proenzymes that are activated under specific conditions. Serine endopeptidases are especially abundant in the plasma, where examples are kallikrein, thrombin and plasmin. There are also large quantities of the proenzymes of trypsin, chymotrypsin and pancreatic elastase in the pancreas, and these are activated when they reach the intestine, by another serine endopeptidase, enteropeptidase.

Serine endopeptidases are also found in the intracellular organelles of neutrophils, mast cells and cytotoxic lymphocytes, and it is those of neutrophils that have been given a great deal of attention in connection with emphysema (Table II).

Table II . Some examples of endopeptidases of the various catalytic types, in the body. Most of the enzymes are synthesised in the form of inactive proenzymes, which require activation; this is indicated for the serine endopeptidases by the inclusion of their recognised pro-enzyme names, but is no less an aspect of the physiology of the metalloendopeptidases, say.

A. Serine Endopeptidases
Pancreatic/intestinal: Trypsin(ogen), chymotrypsin(ogen), pancreatic (pro)elastase
Plasma: (Pro)thrombin and other coagulation enzymes, plasmin(ogen) and plasminogen activators, plasma (pre)kallikrein, complement enzymes
Granulocyte: Neutrophil elastase, cathepsin G, chymase, tryptase, granzymes

B. Cysteine Endopeptidases
Lysosomal: Cathepsin B, cathepsin H, cathepsin L
Cytosolic: Calpains

C. Aspartic Endopeptidases
Stomach: Pepsins
Kidney: Renin
Lysosomes, etc.: Cathepsin D, cathepsin E

D. Metalloendopeptidases
Endogenous: Collagenase (fibroblast and neutrophil types), stromelysin, 72 kilodalton gelatinase, 92 kilodalton gelatinase
Exogenous: Pseudomonas elastase

Unusually for a serine endopeptidase, the elastase of neutrophil leucocytes seems to be stored in the azurophil granules of the cells in its active, processed form, not as a proenzyme.[4] It is prevented from doing damage so long as it is confined in the granule, however. Early work in our own laboratory showed that neutrophil elastase most readily cleaves valyl bonds, whereas the pancreatic elastase acts preferentially on alanyl bonds.[5] Much more has been done on this aspect in the work with synthetic substrates and inhibitors (see Chapter 10). We also showed that the enzyme has very significant activity on the interstitial type I and type II collagens, and also on the collagen of basement membranes. [6,7]

Quantitatively, the major inhibitor of neutrophil elastase in the body is α_1-proteinase inhibitor (see Chapter 5), but in the lung, the bronchial inhibitor may also be important (see Chapter 9).

When elastase is released from the neutrophils, another serine endopeptidase, cathepsin G is also released; this has received much less attention because it has little action on the major substrates, elastin and collagen.

One of the routes for activation of the inactive precursors of the metallo endopeptidases involves the action of a 'cascade', in which plasminogen activators activate plasmin, and the plasmin then initiates the activation of the metalloendopeptidases. This is one of the several ways in which we shall see that the activities of the endopeptidases are interwoven.

Cysteine Endopeptidases

The cysteine endopeptidases are far less numerous in the human body than the serine enzymes. The calcium-dependent calpains are essentially confined inside cells, but the lysosomal cathepsins are secreted from macrophages and tissue cells when they are appropriately stimulated, and therefore may be important in the context of tissue damage.

The activity of the cysteine endopeptidases depends upon cysteine and histidine residues, which permit activity over quite a wide range of acidic to neutral pH. However, the mature forms of lysosomal cathepsins B, H and L are very unstable at pH values of 7 and above, so significant activity is seen only at acidic pH. The cysteine endopeptidases also require reducing conditions for maximal activity, which suits them better for intracellular than extracellular activity.

On the face of it, the requirements of the cathepsins for activity seem to preclude them from extracellular activity, and certainly they cannot be active in the bulk of the extracellular phase. However, careful experiments, over a number of years, with macrophages and osteoclasts, have provided strong evidence that these cells create local regions of acidic pH close to their surfaces, and secrete the lysosomal enzymes into these regions, where they are apparently active. Given the long time-span of development of emphysema, there seems no reason to exclude the possibility that an important component of the degeneration might occur in restricted, pericellular locations around the macrophages of the lung.[8]

The capacity of cathepsins B and L to degrade collagen at acidic pH was established long ago.[9] Much more recently, it was shown that cathepsin L has high activity on elastin, at about pH 5; this activity is as great as that of leucocyte elastase acting at neutral pH.[10] Cathepsin L is also active in the inactivation of α_1-proteinase inhibitor, at slightly acidic pH.[11]

In very recent work with cathepsin B, David Buttle, in Cambridge, with collaborators in several laboratories, has shown that sputum contains a high-molecular mass form of cathepsin B that is stable to neutral pH, like precursor forms of the enzyme recognised previously, but this form is an active endopeptidase (still with an acidic pH optimum).[12] In addition to the effects of pH, the extracellular activity of the cysteine endopeptidases in the body is thought to be controlled by proteins of the cystatin superfamily, which comprise the low molecular mass cystatins and also the kininogens.[13] In extravascular fluids, cystatin C is the major inhibitor, whereas the kininogens are more abundant in the circulation. In a further stage of the collaborative study mentioned above, it has been found that purulent sputum contains a form of cystatin C that has been degraded by leucocyte elastase, and has thus been inactivated[14,15] (D.J. Buttle, personal communication). This graphically illustrates the kind of complex interaction that exists in real-life proteolytic systems: cathepsin L degrades α_1-proteinase inhibitor, tending to release elastase from inhibition, and the elastase degrades cystatin C, so as to relieve the inhibition of the cysteine endopeptidases!

Aspartic Endopeptidases

The best known human aspartic endopeptidases are pepsin, which is largely confined to the stomach, renin, which is involved in the regulation of blood pressure by the kidney, and cathepsin D, which occurs in the lysosomes of cells throughout the body. These enzymes act only in conditions of acidic pH, and have not so far been implicated in lung disease.

Metalloendopeptidases

The polymorphonuclear leucocytes that bring elastase into the lung also bring neutrophil collagenase and the 92 kilodalton form of gelatinase. These metalloendopeptidases act on the interstitial, type I collagen, and basement membrane collagen and other proteins.[16]

When suitably stimulated, as by tobacco smoke or bacteria, alveolar macrophages may be expected to synthesise and secrete both the 92 kilodalton and 72 kilodalton forms of gelatinase, and also stromelysin.[17] These enzymes are capable of making a significant contribution to the degradation of the structural proteins of the lung.

All of the matrix metalloendopeptidases are secreted from cells as their inactive pro-enzyme forms. They are conveniently activated in the test-tube by organomercurial reagents, but the identity of the physiological activators is less clear. As was mentioned previously, plasmin is one of the potential natural activators, and stromelysin is an efficient activator of some of the other enzymes, once it is activated itself.[18]

Tissue fluids contain the so-called "tissue inhibitors of metalloproteinases" or TIMP's, and the metalloendopeptidases are also inhibited by α_2-macroglobulin.

Apart from the numerous endogenous endopeptidases, there is a possibility of degradative enzymes being brought in to the lung by infective organisms. *Pseudomonas aeruginosa*, in particular, produces a powerful elastinolytic endopeptidase. Most bacteria will stimulate the macrophages, and cause infiltration of neutrophils, thus possibly potentiating degradation by endogenous enzymes.

Apart from their direct actions on the structural proteins of the lung, stromelysin and the *Pseudomomas* elastase have both been shown to inactivate α_1-proteinase inhibitor, and thus to have the potential to create a situation resembling the deficiency of this inhibitor, favouring the action of leucocyte elastase. Thus, we conclude this introduction to the endopeptidases with yet another example of the complex ways in which the activities of these enzymes can be interconnected.

References

1. Barrett A.J., McDonald J.K.: Nomenclature: protease, proteinase and peptidase. Biochem. J. 1986; 237:935
2. Barrett A.J.: Introduction: the classification of proteinases. In: Evered D., Whelan J., (Eds.): *Protein Degradation in Health and Disease* (Ciba Foundation Symposium 75), Amsterdam, Excerpta Medica, 1980:1-13
3. Barrett A.J.: A trap for proteinases. Current Contents 1989;32 (2): 14
4. Farley D., Salvesen G., Travis J.: Molecular cloning of human neutrophil elastase. Biol. Chem. Hoppe-Seyler 1988; 369:3-7
5. Blow A.M.J.: Action of human lysosomal elastase on the oxidized B chain of insulin. Biochem. J. 1977;161:13-16
6. Barrett A.J.: The possible role of neutrophilic proteinases in damage to articular cartilage. Agents Actions 1978; 8:11-18
7. Baricos W.H., Zhou Y., Mason R.V., Barrett A.J.: Human kidney cathepsins B and L. Characterization and potential role in glomerular basement membrane degradation. Biochem. J. 1988; 252:301-304
8. Chapman H.A. Jr., Stone O.L.: Co-operation between plasmin and elastase in elastin degradation by intact murine macrophages. Biochem. J. 1984; 222:721-728
9. Kirschke H., Kembhavi A.A., Bohley P., Barrett A.J.: Action of rat liver cathepsin L on collagen and other substrates. Biochem. J. 1982; 201:367-372
10. Mason R.W., Johnson D.A., Barrett A.J., Chapman H.A.: Elastinolytic activity of human cathepsin L. Biochem. J. 1986; 233:925-927
11. Johnson D.A., Barrett A.J., Mason R.W.: Cathepsin L inactivates alpha1-proteinase inhibitor by cleavage in the reactive site region. J. Biol. Chem. 1986; 261:14748-14751
12. Buttle D.J., Bonner B.C., Burnett D., Barrett A.J.: A catalytically active high-M_r form of human cathepsin B from sputum. Biochem. J. 1988; 254:693-699
13. Barrett A.J.: The cystatins: a new class of peptidase inhibitors. Trends Biochem. Sci. 1987; 12:193-196
14. Buttle D.J., Burnett D., Abrahamson M.: Levels of cathepsin B activity and cystatins in human sputum: relationship to inflammation. Scand. J. Clin. Lab. Invest. 1990; 50:509-516

34

15. Abrahamson M., Mason R.W., Hansson H., Grubb A., Buttle D.J. & Ohlsson K.: Leukocyte elastase can inactivate the human cysteine proteinase inhibitor, cystatin C, by cleavage of a single N-terminal bond. Biochem. J. 1990; 273: 621-626
16. Murphy G., Ward R., Hembrey R.M., Reynolds J.J., Kühn K., Tryggvason K.: Characterization of gelatinase from pig polymorphonuclear leucocytes. A metalloproteinase resembling tumour type IV collagenase. Biochem. J. 1989; 258:463-472
17. Senior R.M., Connolly N.L., Cury J.D., Welgus, H.G. , Campbell E.J.: Elastin degradation by human alveolar macrophages. A prominent role of metalloproteinase activity. Am. Rev. Respir. Dis. 1989; 139:1251-1256
18. Nagase H., Enghild J.J. , Salvesen G.: Stepwise activation mechanisms of the precursor of matrix metalloproteinase 3 (stromelysin) by proteinases and (4-aminophenyl) mercuric acetate. Biochemistry 1990; 29:5783-5789

4. Lung Proteinases and Emphysema

J. G. BIETH

INSERM Unit 237, Université Louis Pasteur de Strasbourg, France

Introduction

It is commonly acknowledged that neutrophil elastase (NE) is responsible for lung elastin degradation that leads to emphysema. This theory is based on the following observations:

1. emphysema frequently occurs in α_1-proteinase inhibitor (α_1PI)-deficient individuals,
2. α_1PI is the physiological inhibitor of NE,
3. emphysema-like lesions may be observed after intra-tracheal instillation of NE in the animal.[1]

Patients with hereditary α_1PI deficiency develop panlobular emphysema (destruction of alveoli). This represents only a few percent of all cases of emphysema. The most frequent form of the disease is centrilobular emphysema (destruction of bronchioles) which occurs in α_1PI-sufficient individuals as a result of cigarette smoking. Because these patients have high levels of active α_1PI in their lower respiratory tract,[2] it is hard to believe that neutrophil elastase plays a major role in the pathogenesis of their emphysema. It is likely that the smoker's emphysema is rather a multifactorial disease involving many proteolytic and non-proteolytic factors. This article reviews the lung proteinases that might play a role in the pathogenesis of lung emphysema.

Definition of Elastase

A proteinase may be named "elastase" if it possesses the ability to solubilize mature cross-linked elastin. An enzyme that hydrolyses chemically solubilized elastin or synthetic elastase substrates should not be called elastase if it is unable to solubilize fibrous elastin. Elastases may have variable catalytic sites and widely different primary substrate specificities. They may belong to at least three classes of endopeptidases: serine proteinases (neutrophil elastase), metalloproteinases (macrophage elastase) and cysteine proteinases (macrophage cathepsin L). It must also be emphasized that elastases are not "specific" for elastin: they also hydrolyse other proteins as will be shown below.[1]

Lung Elastases

The elastases present in the lung may originate from neutrophils, macrophages or bacteria. The azurophil granules of polymorphonuclear leukocytes contain three serine proteinases with elastolytic activity: elastase[1], cathepsin G[1] and proteinase 3.[3-4] Human alveolar macrophages are thought to contain three elastolytic proteinases: a metalloproteinase[5], a cysteine proteinase[6] and internalized NE[1](Table I).

Neutrophil elastase

NE is by far the most abundant and the most elastolytic lung elastase. Neutrophils contain large quantities ($3\mu g$ per 10^6 cells) of this powerful enzyme. Extracellular release of NE occurs after cell death, phagocytosis, or neutrophil activation by a variety of stimulants.[7] Smoking induces the *in vivo* release of NE in plasma[8] and

Table I. Neutrophil and macrophage elastases and type-1 helper-proteinases, i.e. enzymes that cleave other lung matrix proteins thus unmasking embedded elastin fibers and rendering their elastolysis easier. Note that some elastases may also act as type-1 helper-proteinases. (PMN = polymorphonuclear neutrophils, MAC = alveolar macrophages)

Elastases	Type-1 Helper-Proteinases
Elastase (PMN,MAC)	Collagenases (PMN, MAC)
Cathepsin G (PMN)	Gelatinases (PMN, MAC)
Proteinase 3 (PMN)	Plasminogen activator (PMN, MAC)
Metallo-elastase (MAC[a], PMN[b])	Elastase (PMN)
Cathepsin L (MAC)	Cathepsin G (PMN)

[a] This is not a single enzyme and should be called "metallo-elastase activity". Unpublished data show that it is a mixture of matrix metalloproteinases 2 and 9 (the 72 kDa and 92 kDa gelatinases) with a small contribution from matrix metalloproteinase 3 (stromelysin) (G. Murphy, *personal communication*).

[b] PMN also contain matrix metalloproteinase 9 (92 kDa gelatinase) (G. Murphy *personal communication*).

in the lower respiratory tract.[9] NE liberated in the lower respiratory tract forms a complex with α_1PI. The levels of this complex are much higher in patients with emphysema[2] than in healthy individuals.[10] NE was found to be associated with elastin fibres in sections of emphysematous lungs.[11]

NE is a basic (pI>9) glycoprotein with a molecular mass of 25-30 kDa. It is composed of a single polypeptide chain of 218 amino acid residues whose sequence is known. The carbohydrates are linked to the peptide chain via Asn 95 and Asn 144.[7] The organization and the chromosomal localization of the elastase gene have been established recently.[12]

The X-ray crystallography of NE complexed to turkey ovomucoid[13] revealed similarities and differences with the model enzyme porcine pancreatic elastase. Neutrophil elastase cleaves preferentially substrates at Val-X or Ala-X bonds. Catalysis is much more efficient with tri- or tetrapeptidic substrates than with short peptides. Succinyl-trialanine-p-nitroanilide and methoxysuccinyl-Ala$_2$-Pro-Val-p-nitroanilide are convenient artificial substrates for NE.

Several human proteins form enzymatically inactive complexes with NE, namely: α_1PI, α_2-macroglobulin, inter-α-inhibitor and mucus proteinase inhibitor. The physiological NE inhibitor of the lower respiratory tract is α_1PI whose concentration in the epithelial lining fluid is 4.6 μM in individuals with the normal PiM$_1$M$_1$ phenotype, 0.46 μM in PiZZ α_1PI deficient patients[14], and null in Pi Nul Nul subjects.[15] α_1PI is an irreversible NE inhibitor whose high k$_{ass}$ (10^7 M^{-1} s^{-1}) corresponds to a delay time of inhibition of 0.12 seconds in the lower respiratory tract of healthy individuals.[14] The delay time of inhibition is the time required to almost fully inhibit elastase $in\ vivo$.[16] The physiological elastase inhibitor of the upper respiratory tract is mucus proteinase inhibitor, a reversible inhibitor with k$_{ass}$ = 6x10^6 M^{-1} s^{-1}, k$_{diss}$ = 2.3x10^{-3} s^{-1} and Ki=3x10^{-10} M.[17] The molar ratio of mucus proteinase inhibitor to α_1PI in the lower respiratory tract is about 0.1.[2]

The two other inhibitors cited above are unimportant physiologically: α_2-macroglobulin occurs in very low amounts in the lung and inter-α-inhibitor forms a loose complex with elastase.[18]

Neutrophil cathepsin G

Cathepsin G also exists in high amounts in neutrophils (1-2 μg per 10^6 cells) and shares many physical-chemical properties with NE: both are basic glycoproteins, occur as isoenzymes, are composed of a single polypeptide chain, and have comparable molecular masses. The structure of cathepsin G was determined recently by analysis of a cDNA coding for this enzyme.[19] The enzyme shares less than 36% sequence identity with neutrophil elastase. The genomic organization and chromosomal localization of the cathepsin G gene have been established.[20]

The elastolytic activity of cathepsin G is about 10% that of NE. In addition, cathepsin G enhances the elastolytic activity of elastase.[21-22] Cathepsin G is a chymotrypsin-like serine proteinase which may be conveniently assayed with succinyl-Ala$_2$-Pro-Phe-p-nitroanilide.

Among the endogenous inhibitors of cathepsin G (α_1PI, α_1-antichymotrypsin, α_2-macroglubulin, mucus proteinase inhibitor), α_1-antichymotrypsin has the highest reaction rate ($k_{ass} = 3 \times 10^7 \, M^{-1} \, s^{-1}$).[1] This inhibitor occurs, however, at a relatively low concentration in plasma and hence in the epithelial lining fluid. The activity of cathepsin G in the lower respiratory tract is therefore likely to be controlled by α_1PI ($k_{ass} = 4 \times 10^5 \, M^{-1} \, s^{-1}$, corresponding to a delay time of inhibition of about 3 seconds[1]). On the other hand, mucus proteinase inhibitor is probably the physiological cathepsin G inhibitor in the upper respiratory tract.

Neutrophil proteinase 3

Proteinase 3 is a serine proteinase able to solubilize fibrous elastin and to induce emphysema in the hamster.[3] Its elastolytic activity is slightly higher than that of NE at pH 6.5 but is significantly lower than NE at pH7.4. Proteinase 3 does not hydrolyze synthetic substrates of elastase and cathepsin G. It is significantly less basic (pI = 9.1) than these two enzymes. The quantity of proteinase 3 purified from neutrophil granules is about on third that of NE. No natural inhibitor is known for this enzyme. Preliminary data from our laboratory show that α_1PI and mucus proteinase inhibitor do not inhibit proteinase 3.

Macropages' neutrophil elastase

Human alveolar macrophages have been shown to internalize human NE *in vitro* and *in vivo* and to release this enzyme in an active form.[1] By this mechanism NE may escape the action of endogenous elastase inhibitors and may be liberated at sites where the antielastase protection is low.

Macrophage metalloproteinase with elastase activity

This is an ill-defined enzyme activity.[1] Senior et al[5] recently studied elastin degradation by human alveolar macrophages cultured on elastin. They found elastin degradation in this model system. The elastolytic activity was inhibited by tissue inhibitor of metalloproteinases (TIMP), suggesting that it is due to a metalloproteinase. The TIMP are a family of proteins able to inhibit metalloproteinases such as collagenases, gelatinases and stromelysin.[23] See also footnote "a" to Table I.

Macrophage cathepsin L

Human alveolar macrophages have been shown to degrade elastin by a process that requires cell-substrate contact and involves acidic cysteine proteinases.[1] Liver lysosomal cathepsin L, a cysteine proteinase, solubilizes elastin at a slightly acidic pH. An immunologically and functionally related enzyme has been partially purified from human alveolar macrophages. Macrophages are able to generate a local acidic environment. The participation of macrophage cathepsin L in the degradation of elastin is therefore possible (ref.[6] and refs therein). Plasma derived cysteine proteinase inhibitors might control this activity.[24]

Bacterial elastases

Elastolytic activity has been found in a large number of bacteria.[1] Among others, *Pseudomonas aeruginosa* which is frequently involved in lung infections, contains a powerful Zn^{2+} metalloelastase which solubilizes human lung elastin rapidly.[25] A convenient and commercially available fluorogenic substrate of this proteinase is 2-aminobenzoyl-Ala-Gly-Leu-Ala-4-nitrobenzylamide.[1] *P. aeruginosa* elastase is inhibited by α_2-macroglobulin but not by TIMP.

Lung Helper-Proteinases

Helper-proteinases are defined as non-elastolytic enzymes able to potentiate the action of true elastases through two different mechanisms:
- cleavage of extracellular matrix proteins which unmasks embedded elastin fibres and eases their elastase-catalyzed breakdown (type-1 helper-proteinases) (Table I);
- inactivation of elastase inhibitors and other proteinase inhibitors which allows unimpaired proteolysis (type-2 helper-proteinases) (Table II).

Type-1 helper-proteinases

Lung elastin fibres are embedded in a complex mix of collagen, proteoglycans and glycoproteins. Degradation of these matrix components probably facilitates the action of elastases on elastin. NE and cathepsin G themselves are able to cleave interstitial collagen, proteoglycans and fibronectin.[1]

Neutrophils and macrophages synthesize latent collagenases named matrix metalloproteinase 9 (neutrophils) and matrix metalloproteinase 1 (macrophages). These enzymes are active on lung interstitial (type I, II and III) collagens. Both cells also synthesize a latent gelatinase (matrix metalloproteinase 2) acting on degraded interstitial collagen and on type V collagen. The neutrophils' collagenase is stored in the specific granules of these cells. The sequence of its 467 aminoacids has recently been deduced from the nucleotide sequence of a cDNA clone.[26] Macrophages also secrete matrix metalloproteinase 3 also called stromelysin or proteoglycanase which also acts on matrix components. All three matrix metalloproteinases are secreted as latent enzymes. They may be activated *in vitro* with organomercurials or trypsin. *In vivo* activation may take place with plasmin, NE and cathepsin G[27] or with oxidants.[27] The physiological inhibitors of these proteinases are α_2-macroglobulin and TIMP.[28-29]

The specific granules of neutrophils contain plasminogen activator, a trypsin-like proteinase that is also associated with the cell membrane of macrophages. This

Table II. Type-2 helper-proteinases i.e. enzymes that inactivate protein proteinase inhibitors through proteolytic cleavage

Inhibitor	Target proteinases	Type-2 helper-proteinases
α_1-proteinase inhibitor	neutrophil elastase (neutrophil cathepsin G)	neutrophil metalloproteinase; macrophage cathepsin L; *Serratia marcescens* metalloproteinase; *Pseudomonas aeruginosa* elastase; *Staphylococcus aureus* proteinases; *Legionella pneumophila* proteinase
α_1-antichymotrypsin	neutrophil cathepsin G	neutrophil elastase; neutrophil collagenase
α_2-antiplasmin	plasmin	neutrophil elastase
tissue inhibitor of metalloproteinases (TIMP)	matrix metalloproteinase (collagenase, gelatinase, stromelysin)	neutrophil elastase
cystatin C	cathepsin L	neutrophil elastase[a]

[a] See Chapter 3

enzyme generates plasmin from plasminogen, a pro-enzyme present in the lung interstitium.[30] Plasmin activates latent matrix metalloproteinases and cleaves fibronectin, laminin and other interstitial glycoproteins. Its activity is regulated by α_2-macroglobulin and α_2-antiplasmin. The physiological inhibitor of plasminogen activator is plasminogen activator inhibitor 1 whose inhibitor properties are impaired by oxidants.[28,29] Chapman et al.[30] recently cultured macrophages on an elastin-rich extracellular matrix and showed that the rate of elastin solubilization markedly increased upon addition of plasminogen thus providing *in vitro* evidence for the role of plasminogen activator as a helper-proteinase.

Type-2 helper-proteinases

The activity of most elastases and type-1 helper-proteinases is normally controlled by protein proteinase inhibitors. The latter may be proteolytically inactivated by type-2 helper-proteinases (Table II). α_1PI is inactivated by phagocyte [31-33] and bacterial proteinases.[34-37] Table III shows that the inactivation of α_1PI occurs as a result of the proteolytic cleavage of the inhibitor at or near its active centre (Met$_{358}$-Ser$_{359}$). The anti-NE shield of the lower respiratory tract may therefore be weakened during phagocyte activation and lung infections.

On the other hand, NE is able to inactivate α_1-antichymotrypsin[38], α_2-antiplasmin[39] and TIMP.[40] This will facilitate the action of type-1 helper-proteinases on lung matrix proteins.

Non-Proteolytic Helper-Systems

A number of non-proteolytic factors may decrease the functional activity of lung elastase inhibitors and/or increase the level of lung elastolytic and non-elastolytic proteinases.

Oxidants

Cigarette smoke-activated phagocytes secrete oxidants that impair the activity of α_1PI. Cigarette smoke itself is able to do so. The inactivation takes place through oxidation of methionine 358 in the active centre of α_1PI into methionine sulphoxide.[41] These discoveries led to the hypothesis that the smokers' emphysema may occur as a result of an acquired α_1PI defect in activity comparable to the genetic defect of the protein. This theory no longer holds true. Healthy smokers and nonsmokers have similar levels of active bronchoalveolar lavage α_1PI.[42-43] In addition, α_1PI-sufficient patients with emphysema have levels of active α_1PI similar to those of age-matched patients with sarcoidosis.[2] Further, Campbell et al.[44] designed an ELISA specific for oxidized α_1PI and showed that the concentration

Table III. Proteolytic inactivation of α_1-proteinase inhibitor by type-2 helper-proteinases. Limited proteolysis at the bait region of the inhibitor

Residue n.	352	353	354	355	356	357	358	359
Sequence	Phe -	Leu -	Glu -	Ala -	Ile -	Pro -	Met -	Ser
		↑		↑			↑	↑
		4		3			2	1

1 *Serratia marcescens* metalloproteinase, cathepsin L
2 *Pseudomonas aeruginosa* elastase
3 *Staphylococcus aureus* serine and cysteine proteinases, cathepsin L
4 *Staphylococcus aureus* metalloproteinase, neutrophil metalloproteinase

of the oxidized inhibitor in bronchoalveolar lavage fluids accounted for less than 0.1% of the total α_1PI. Also, smokers and nonsmokers had identical levels of oxidized α_1PI. Smoking therefore does not lead to a significant decrease of functional α_1PI as thought previously. Local decreases obscured by the dilution due to the lung lavage procedure may, however, take place.

Oxidants may have at least three other deleterious effects (Fig.1). They are able to activate latent neutrophil metalloproteinase.[28] As discussed above, active metalloproteinases inactivate α_1PI (Tables II,III). Oxidants inactivate plasminogen activator inhibitor I[45] thus leading to an increased generation of plasmin which, in turn, activates latent metalloproteinases. Last, oxidants decrease the activity of lysyloxidase thus impairing the resynthesis of elastin.[41]

Other factors

Chemoattractants obviously play an important role in emphysema as they recruit phagocytes in the lung.[29] It is worthwhile noticing that proteolytically inactivated α_1PI as well as the α_1PI-neutrophil elastase complex are also neutrophil chemoattracts.[46-47] Last, let us mention two substances that might modulate the *in vivo* activity of neutrophil elastase. Platelet factor 4 and other lysine-rich ligands stimulate the elastolytic activity of neutrophil elastase.[48] On the other hand, heparin and other glycosaminoglycans decrease the rate of inhibition of neutrophil elastase by α_1PI.[49]

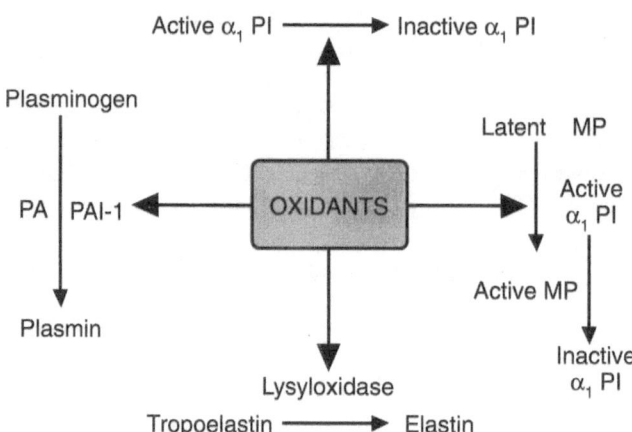

Fig. 1. Possible effects of oxidants on the elastase/inhibitor/elastin system. (α_1PI = α_1-proteinase inhibitor, PA = plasminogen activator, MP = metalloproteinase, PAI-1=plasminogen activator inhibitor 1).

Conclusion

In addition to the well-characterized NE, lung phagocytes contain at least four other elastases whose importance is frequently overlooked. The lung also contains helper-proteinases some of which cleave matrix proteins other than elastin, thus facilitating the action of elastase while others degrade proteinase inhibitors. A member of non-proteolytic helper systems including oxidants may also play an important role in emphysema.

The numerous pathogenic factors reviewed here are obviously difficult to sort out. Panlobular emphysema due to a genetic defect of α_1PI is probably the most easy form of emphysema to rationalize: α_1PI, the specific NE inhibitor, is absent (PiNullNull individuals) or exist in low amounts (PiZZ patients) so that elastase may attack elastin fibres in an unimpaired way. The pathogenic role of NE in this type of disease appears to be clearcut which fully justifies the ongoing α_1PI replacement therapy[50] or the design of synthetic elastase inhibitors as antiemphysema drugs.[51]

Centrilobular emphysema, the smokers' disease, is more difficult to rationalize. Here α_1PI is present and functionally active in the lung which makes it hard to believe that NE can be an important pathogenic factor.

We have suggested earlier[1] a scenario in which NE and other neutrophil proteinases serve mainly to "clean" the elastic fibres from their surrounding connective tissue macromolecules. Elastolysis will then be achieved by the three elastases of the resident macrophage through close cell-matrix contacts which prevent the action of the surrounding inhibitors. *In vitro* studies show that macrophage-elastin contacts lead to efficient elastolysis by cathepsin L[30] or metallo-elastase[5] in the presence of inhibitors.

Smokers have many more macrophages in their lower respiratory tract than nonsmokers. In addition, a markedly increased number of these cells are found in the vicinity of those small airways where the destructive changes are the most prominent.[52]

Whatever the scenario, it is very likely that the smoker's emphysema is a multifactorial disease. Drug designers should keep this in mind and perhaps think about making cysteine and metalloproteinase inhibitors in addition to neutrophil elastase inhibitors.

Acknowledgments

The Author thanks Dr. A.J. Barret, Strangeways Research Laboratory, Cambridge, England for carefully reading and improving the manuscript

References

1. Bieth J.G.: Elastases: catalytic and biological properties. In: Mecham R.P. (Ed.) *Regulation of*

matrix accumulation. New York, Academic Press, 1986; 217-320

2. Gast A., Dietemann-Molard A., Pelletier A., Pauli G., Bieth J.G.: The antielastase screen of the lower respiratory tract of α_1-proteinase inhibitor-sufficient patients with emphysema or pneumothorax. Am. Rev. Respir. Dis. 1990; 141:880-883

3. Kao R.C., Whener N.G., Skubitz K.M., Gray B.H., Hoidal J.R.: Proteinase 3. A distinct human polymorphonuclear leukocyte proteinase that produces emphysema in hamsters. J. Clin. Invest. 1988; 82:1963-1973

4. Goldschmeding R., van der Schoot C.E., ten Bokkel Huinink D., Hack C.E., van den Ende M.E., Kallenberg C.G.M., von dem Borne A.E.G.Kr.: Wegener's granulomatosis autoantibodies identify a novel diisopropylfluorophosphate-binding protein in the lysosomes of normal human neutrophils. J. Clin. Invest. 1989; 84:1557-1587

5. Senior R.M., Connoly N.L., Cury J.D., Welgus H.G., Campbell E.J.: Elastin degradation by human alveolar macrophages. Am. Rev. Respir. Dis. 1989; 139:1251-1256

6. Reilly J.J. Jr, Mason R.W., Chen P., Joseph L.J., Sukhatme V.P., Yee R., Chapmann H.A. Jr.: Synthesis and processing of cathepsin L, and elastase, by human alveolar macrophages. Biochem. J. 1989; 257:493-498

7. Bieth J.G.: Human neutrophil elastase. In: Robert L., Hornebeck W. (Eds.) *Elastin and elastases.* Boca Raton, CRC Press, 1989; vol II, 24-31

8. Weitz J.I., Crowley K.A., Landman S.L., Lipman B.I., Yu J.: Increased neutrophil elastase activity in cigarette smokers. Ann. Intern. Med. 1987; 107:680-682

9. Abbound R., Fera T., Ritcher A., Tabona M.Z., Johal S.Z.: Acute effect of smoking on the functional activity of alpha-1-proteinase inhibitor in bronchoalveolar lavage fluid. Am. Rev. Respir. Dis. 1985; 131:79-85

10. Jochum M., Pelletier A., Boudier C., Pauli G., Bieth J.G.: The concentration of leukocyte elastase α_1-proteinase inhibitor complex in bronchoalveolar lavage fluids from healthy human individuals. Am. Rev. Respir. Dis. 1985; 132:913-914

11. Damiano V.V., Tsang A., Kucich U., Abrams W.R., Rosenbloom J. Kimbel P., Fallahnejad M., Weinbaum G.: Immunolocalization of elastase in human emphysematous lungs. J. Clin. Invest., 1986; 78:482-493

12. Takahashi H., Nukiwa T., Yoshimura K., Quick C.D., States D.J., Holmes M.D., Whang-Peng J., Knutsen T., Crystal R.G:: Structure of the human neutrophil elastase gene. J. Biol. Chem. 1988; 263:14739-14747

13. Bode W., An-Zhi W., Huber R., Meyer E., Travis J., Neumann S.: X-ray crystal structure of the complex of human leucocyte elastase (PMN elastase) and the third domain of the turkey ovomucoid inhibitor. EMBO J. 1986; 5:2453-2458

14. Ogushi F., Hubbard R.C., Fells G.A., Casolaro M.A., Curiel D.T., Brantly M.L., Crystal R.G.: Evaluation of the S-type of alpha-1-antitrypsin as an *in vivo* and *in vitro* inhibitor of neutrophil elastase. Am. Rev. Respir. Dis. 1988; 137:364-370

15. Cox D.W., Levison H.: Emphysema of early onset associated with a complete deficiency of alpha-1-antitrypsin (null homozygotes). Am. Rev. Respir. Dis. 1988; 137:371-375

16. Bieth J.G.: Pathophysiological interpretation of kinetic constants of protease inhibitors. Bull. Europ. Physiopath. Respir. 1980; 16(suppl.):183-195

17. Boudier C., Bieth J.G.: Mucus proteinase inhibitor: a fast-acting inhibitor of leucocyte elastase. Biochim. Biophys. Acta. 1989; 995:36-41

18. Gast A., Bieth J.G.: Inhibition of human neutrophil elastase by acid soluble inter-α-trypsin inhibitor. Adv. Exp. Med. Biol. 1988; 240:75-82

19. Salvesen G., Farley D., Shuman J., Przybyla A., Reilly C., Travis J.: Molecular cloning of human cathepsin G: structural similarity to mast cell and cytotoxic T lymphocyte proteinase. Biochem-

istry 1987; 26:2289-2293

20. Holn P.A., Popescu N.C., Hanson R.D., Salvesen G., Ley T.J.: Genomic organization and chromosomal localization of the human cathepsin G gene. J. Biol. Chem. 1989; 264:13412-13419

21. Boudier C., Holle C., Bieth J.G.: Stimulation of the elastolytic activity of leukocyte elastase by leukocyte cathepsin G. J. Biol. Chem. 1981; 256:10256-10258

22. Lucey E.C., Stone P.J., Breurer R., Christensen T.G., Calore J.D., Catanese A., Franzblau C., Snider G.L.: Effects of combined human neutrophil cathepsin G and elastase on induction of secretory cell metaplasia and emphysema in hamsters with *in vitro* observations on elastolysis by these enzymes. Am. Rev. Respir. Dis. 1985; 132:362-366

23. Cawston T.E.: Protein inhibitors of metallo-proteinases. In: Barrett A.J., Salvesen G. (Eds.) *Proteinases inhibitors*. New York, Elsevier, 1986; 589-610

24. Barrett A.J., Rawlings D., Davies M.E., Machleidt W., Salvesen G., Turk V.: Cysteine proteinase inhibitors of the cystatin superfamily. In: Barrett A.J., Salvesen G. (Eds.) *Proteinases inhibitors*. New York, Elsevier, 1986; 515-569

25. Hamdaoui A., Wund-Bisseret F., Bieth J.G.: Fast-solubilization of human lung elastin by *Pseudomonas aeruginosa* elastase. Am. Rev. Respir. Dis. 1987; 135:860-863

26. Hasty K.A., Pourmotabbed T.F., Goldberg G.I., Thompson J.P., Spinella D.G., Stevens R.M., Mainardi C.L.: Human neutrophil collagenase. A distinct gene product with homology to other matrix metalloproteinases. J. Biol. Chem. 1990; 265: 11421-11424

27. Okada Y., Nakanishi I.: Activation of matrix metalloproteinase 3 (stromelysin) and matrix metalloproteinase 2 ("gelatinase") by human elastase and cathepsin G. FEBS Lett. 1989; 249: 353-356

28. Weiss S.J.: Mechanisms of disease. Tissue destruction by neutrophils. N. Engl. J. Med. 1989; 320:365-376

29. Sibille Y., Reynolds H.Y.: Macrophages and polymorphonuclear neutrophils in lung defence and injury. Am. Rev. Respir. Dis. 1990; 141:471-501

30. Chapman H.A. Jr., Reilly J.J. Jr., Kobzik L.: Role of plasminogen activator in degradation of extracellular matrix protein by live human alveolar macrophages. Am. Rev. Respir. Dis. 1988; 137:412-419

31. Desrochers P.E., Weiss S.: Proteolytic inactivation of alpha-1-proteinase inhibitor by a neutrophil metalloproteinase. J. Clin. Invest. 1988; 81:1646-1650

32. Vissers M.C.M., George P. M., Bathurt I.C., Brennan O.S., Winterbourn C.C.: Cleavage and inactivation of α_1-antitrypsin by metalloproteinases released from neutrophils. J. Clin. Invest. 1988; 82:706-711

33. Johnson D.A., Barrett A.J., Mason R.W.: Cathepsin L inactivates α_1-proteinase inhibitor by cleavage in the reactive site region. J. Biol. Chem. 1986; 261:14748-14751

34. Virca G.D., Lyerly D., Kreger A., Travis J.: Inactivation of α_1-proteinase inhibitor by *Serratia marcescens* metalloproteinase. Biochim. Biophys. Acta 1982; 704:267-271

35. Morihara K., Tsuzuki H., Harada M., Iwata T.: Purification of human plasma α_1-proteinase inhibitor and its inactivation by *Pseudomonas* elastase. J. Biochem. 1984;95:795-804

36. Potempa J. Watorek W., Travis J.: Inactivation of human α_1-proteinase inhibitor by *Staphylococcus aureus* proteinases. J. Biol. Chem. 1986;261:14330-14334

37. Conlan J.W., Williams A., Ashworth L.A.E.: Inactivation of human α_1-antitrypsin by a tissue-destructive protease of *Legionella pneumophila*. J. Gen. Microbiol. 1988; 134:481-487

38. Morii M., Travis J.: Amino acid sequence at the reactive site of human α_1-antichymotrypsin. J. Biol. Chem. 1983;258:12749-12752

39. Brower M.S., Harpel P.C.: Proteolytic cleavage and inactivation of α_2-plasmin inhibitor and C1 inactivator by human polymorphonuclear leukocyte elastase. J. Biol. Chem. 1982;257:9854-

46

9859

40. Okada Y., Watanabe S., Nakanishi I., Kishi J.I., Hayakawa T., Watorek W., Travis J., Nagase I.: Inactivation of tissue inhibitor of metalloproteinases by neutrophil elastase and other proteinases. FEBS Lett. 1988;229:157-160

41. Janoff A.: Emphysema: proteinase-antiproteinase imbalance. In Gallin J.I., Goldstein I.M, Snyderman R. (eds): *Inflammation: basic principles and clinical correlates.* New York, Raven Press, 1988; 803-814

42. Boudier C., Pelletier A., Pauli G., Bieth J.G.: The functional activity of α_1-proteinase inhibitor in bronchoalveolar lavage fluids from healthy human smokers and non-smokers. Clin. Chim. Acta. 1983;132:309-315

43. Stone P.J., Calore J.D.: Functional α_1-proteinase inhibitor in the lower respiratory tract of cigarette smokers is not decreased. Science 1983;221:1187-1189

44. Campbell E.J., Endicott S.K., Rios-Mollineda R.A.: Assessment of oxidation of α_1-proteinase inhibitor in broncho-alveolar lining fluid by monoclonal immunoassay: comparison of smokers and non-smokers. Am. Rev. Respir. Dis. 1987;135: A156

45. Lawrence D.A., Loskutoff D.J.: Inactivation of plasminogen activator inhibitor by oxidants. Biochemistry 1986;25:6351-6355

46. Banda M.J., Rice A.G., Griffin G.L., Senior R.M.: α_1-proteinase inhibitor is a neutrophil chemoattractant after proteolytic inactivation by macrophage elastase. J. Biol. Chem. 1988;263:4481-4484

47. Banda M.J., Rice A.G., Griffin G.L., Senior R.M.: The inhibitory complex of human α_1-proteinase inhibitor and human leukocyte elastase is a neutrophil chemoattractant. J. Exp. Med. 1988;167:1608-1615

48. Lonky S.A., Wohl H.: Stimulation of human leukocyte elastase by platelet factor 4. Physiologic, morphologic and biochemical effects on hamster lungs in vitro. J. Clin. Invest. 1981;67:817-826

49. Frommherz K., Bieth J.G.: Heparin decreases the rate constant for the inhibition of leucocyte elastase by α_1-proteinase inhibitor. Eur. Resp. J. 1989;2 suppl. 5:347S

50. Hubbard R.G., Sellers S., Czerski D., Stephens L., Crystal R.G.: Biochemical efficacity and safety of monthly augmentation therapy for α_1 antitrypsin deficiency. JAMA 1988; 260:1259-1264

51. Fletcher D.S., Osinga D.G., Hand K.M., Dellea P.S., Ashe B.M., Mumford R.A., Davies P., Hagmann W., Finke P.E., Doherty J.B., Bonney R.J.: A comparison of α_1-proteinase inhibitor methoxysuccinyl-Ala-Ala-Pro-Val-Chloromethylketone and specific β-lactam inhibitors in an acute model of human polymorphonuclear leukocyte elastase-induced lung hemorrhage in the hamster. Am. Rev. Respir. Dis. 1990;141:672-677

52. Niewoehner D.E., Kleinerman J., Rice D.B.: Pathologic changes in the peripheral airways of young cigarette smokers. N. Engl. J. Med. 1974;291:755-758

5. Proteinases and Proteinase Inhibitors in the Pathogenesis of Pulmonary Emphysema in Humans

R. A. Stockley, D. Burnett

Lung Immunobiochemical Research Laboratory, The General Hospital, Birmingham, UK

Central to our concepts of the pathogenesis of emphysema is the assumption that degradation of lung elastin occurs due to an imbalance between enzymes with the ability to digest this connective tissue and the inhibitors which protect it. This assumption forms the basis of the proteinase/antiproteinase theory of the pathogenesis of emphysema which has dominated the field of emphysema research for more than 25 years.

The proteinase/antiproteinase hypothesis originated with the observation that subjects with inherited deficiency of plasma α_1 antitrypsin (α_1AT) were particularly susceptible to the development of emphysema.[1] At the same time animal studies showed that a proteolytic enzyme was able to destroy lung tissue, leading to emphysema.[2] Since α_1AT is a major plasma inhibitor of serine proteinases[3] it was suggested that subjects with α_1AT deficiency developed emphysema because they had little ability to protect the lung from damage by these enzymes.

Subsequent studies showed that only enzymes which possessed elastolytic properties were capable of producing emphysema experimentally in animals and the human enzyme, neutrophil elastase (NE), became the obvious candidate. This possibility was supported by animal studies confirming that NE could cause emphysema[4] and the fact that α_1AT is the major plasma inhibitor of NE. Thus research into human emphysema has concentrated (almost exclusively) on the role of α_1AT and NE in the pathogenesis of the disease.

The proteinase/antiproteinase theory of the pathogenesis of emphysema has assumed that under certain circumstances the release of NE in the lung is sufficient to overcome the ability of α_1AT to inhibit the enzyme completely with the consequence that free enzyme activity persists. When this occurs the NE causes degradation of lung elastin which, if persistent or recurrent over many years, will lead to clinically important emphysema. If this simple assumption is correct any factor which leads to a disturbance of the enzyme/inhibitor balance in favour of the enzymes would cause emphysema. Such an imbalance could occur for one of three possible reasons:

a. a defective inhibitor screen;
b. an increased enzyme load;
c. a combination of both.

The purpose of this review is to explore the evidence to support or refute this concept, as well as indicate other possible mechanisms for future study.

Role of α_1AT

Alpha$_1$AT has been considered to be the major determinant in the evolution of emphysema. Certainly subjects with plasma deficiency are particularly susceptible to the development of the disease and the concentrations of α_1AT and its function in lung fluids of these subjects are undoubtedly reduced.[5,6] Nevertheless, lung α_1AT still possesses some elastase inhibitory capacity even in subjects where emphysema is present and free elastase activity is not detected.[6] Superficially this would appear to invalidate the proteinase/antiproteinase theory of emphysema. However such studies on lung lavage, though often cited, are difficult to interpret for several reasons:

1. The proteinase/antiproteinase balance may be disturbed on an intermittent basis rather than continuously and this may be difficult to detect on a single occasion.
2. Lavage fluid is harvested from a large area of the lung and may not reflect small localised pockets of imbalance and disease.
3. The disease process occurs in the interstitial space where the lung elastin is localised. Thus changes in lavage fluid may not reflect changes in the local environment where elastin degradation is thought to occur. (This point will be covered in more detail later).
4. The presence of α_1AT deficiency alone does not inevitably lead to the development of emphysema. Smoking undoubtedly increases the susceptibility of these individuals to develop disease,[7] although even some smokers may retain relatively normal lung function until old age.[8] This suggests that factors other than α_1AT deficiency are also important. Indeed it is becoming apparent that subjects, detected by random screening, remain healthy[9] and it has been recognised for years that other relatives of α_1AT deficiency subjects are also sus-

ceptible to the development of emphysema even if they have normal plasma α_1AT concentrations.[10]

Thus it seems likely that other factors are important in causing emphysema even in subjects with α_1AT deficiency. In this respect the presence of large numbers of neutrophils in the lung lavage fluid[6] may play an additional role. Mechanisms causing this accumulation of cells are unknown although a factor which controls neutrophil migration has also been found to be deficient in subjects with α_1AT deficiency[11] and this phenomenon clearly warrants further study.

The central role of α_1AT in the pathogenesis of emphysema becomes more complicated and more tenuous in subjects who appear to have normal plasma and hence lung α_1AT. Such subjects comprise the vast majority (>95%) of the patients with emphysema. In these subjects it has been suggested that it is not the amount, but the function of lung α_1AT that determines the susceptibility to emphysema. For this reason factors which reduce the function of lung α_1AT could play a major role by decreasing the anti-elastase screen and thus making it easier for NE released in the lung to overcome the protective screen and damage lung elastin.

The function of α_1AT can be reduced by several mechanisms:

1. Oxidation of the active site methionine residue by superoxides from cigarette smoke or inflammatory cells results in a 2,000-fold decrease in the ability of α_1AT to inhibit NE.
2. Cleavage of the α_1AT at or near the active site leads to complete inactivation of the inhibitor and release of a 4,000 Da carboxy terminal peptide.
3. Irreversible complexing with proteinases including NE resulting in a combined protein of 80,000 Da causes inactivation of the inhibitor.
4. Non-oxidative inactivation by cigarette smoke has also been proposed.[12]

Early evidence suggested that the inhibitory function of α_1AT was indeed diminished in the lung lavage of smokers[13] and it was suggested that this may represent oxidation of the active site methionine by cigarette smoke. Although the result of this study was supported by several others[14,15], the relevance to the pathogenesis of emphysema remained uncertain. This doubt was based on two observations. Firstly, the α_1AT remained sufficiently active to inhibit NE although again it was argued that the data reflected an average result from several areas of the lung and some areas could contain totally inactive α_1AT.

Secondly, and more difficult to explain, was the fact that the results were obtained from healthy smokers and it is well known that only 10-15% of these subjects would eventually develop a degree of emphysema which is clinically important. Thus the results from a large group of normal subjects may conceal the most important data from the few individuals who are destined to develop emphysema. No attempt has yet been made to identify these "outliers".

The relevance of the findings of these 3 studies [13-15] to the pathogenesis of emphysema has been offset by subsequent studies that have failed to demonstrate a reduction in α_1AT function in the lungs of healthy smokers compared to non-smokers.[16-18]

Furthermore, studies of α_1AT function in patients with emphysema who continue to or have ceased smoking indicates no difference.[19] It is well known that cessation of smoking leads to a reduction in the progression of emphysema. Thus if modification of α_1AT function by cigarette smoke is a major determinant of progression in emphysema, patients who have stopped smoking should have greater α_1AT function in the lung. Thus the clinical studies have failed to provide clear evidence for the central role of α_1AT in the pathogenesis of emphysema.

Nevertheless these latter studies[16-19] have shown a reduction in the inhibitory capacity of α_1AT. In some studies oxidation of the active site has been suggested as the explanation although few direct experiments have been performed.

Carp et al.[14] demonstrated the presence of sulphoxidised methionine residues in the α_1AT of lung lavage fluids from smokers but not non-smokers. However Campbell and colleagues have been able to detect significant quantities of oxidised α_1AT (using a specific monoclonal antibody) in the lung lavage from healthy smokers.[20] The only other evidence for oxidised α_1AT in lung secretions comes from our own studies on sputum from patients with established emphysema.[21] The study indicated that a significant degree of α_1AT inhibitory function could be restored following incubation with the enzyme methionine sulphoxide peptide reductase, which reverses the inactivation of α_1AT by reduction of the oxidised methionine residue.[22] However, when similar techniques were applied to lavage fluids from normal smokers and non-smokers no real increase in α_1AT function was observed.[18] These data would question the role of oxidation of α_1AT in the pathogenesis of emphysema. Whether such a mechanism could have relevance to the immediate local environment around the inflammatory cells in the lung tissues will be addressed later.

Other methods of α_1AT inactivation have also been demonstrated. Polyacrylamide gel electrophoresis has shown the presence of α_1AT species of various molecular sizes, suggesting complex formation with enzymes as well as limited cleavage.[19] However the nature of the enzymes involved is largely unknown, although some α_1AT/NE complex has been found in lung secretions in several studies.[23,24]

Despite all of these studies and an increase in our knowledge of the nature of lung α_1AT, no clear message has emerged which is unique to patients with emphysema or the small proportion of smokers who are likely to be at risk. However more recent genetic studies may have provided a further clue. Several polymorphisms of the α_1AT gene have been described and, of these, one associated with a restriction site

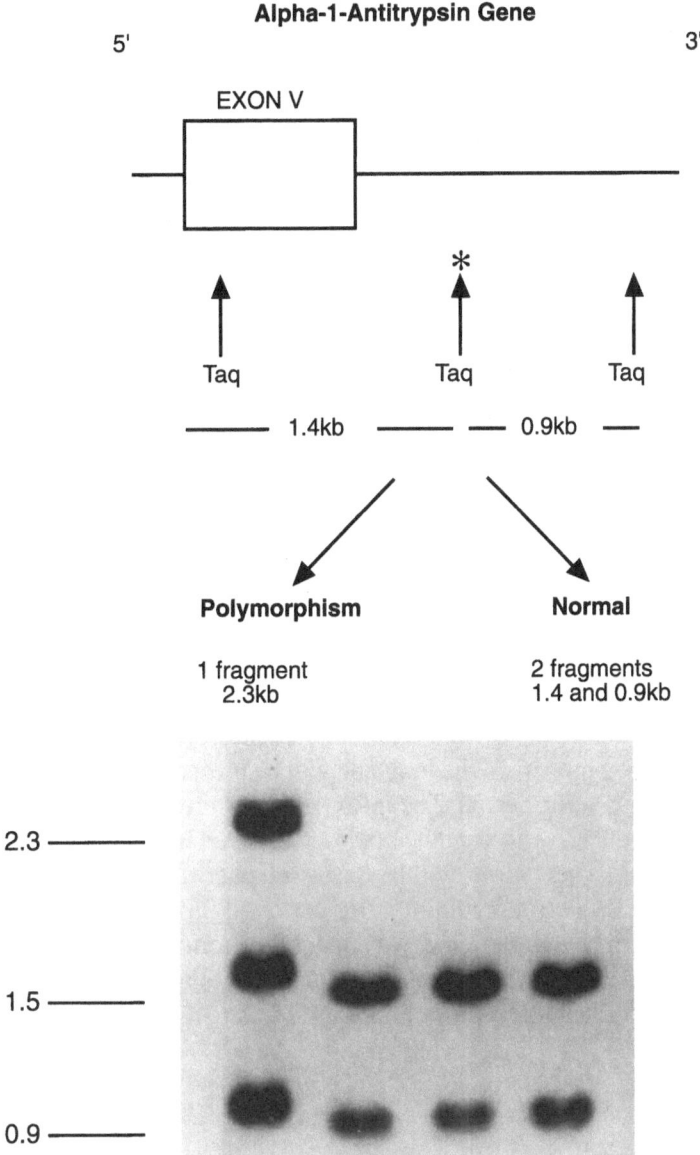

Fig. 1. Diagram to indicate 3 Taq 1 restriction sites at the 3' end of the α_1AT gene. One is in exon V and 2 in the 3' flanking region. The middle site is the one affected by the polymorphism. Digestion by the enzyme and separation by gel electrophoresis produces one or two fragments depending on the presence of the polymorphism and identified by a radiolabelled probe complementary to this region of the α_1AT gene. The gel indicates one subject with the polymorphism and 3 normal subjects. Three bands are seen for the subject with the polymorphism indicating a heterozygous state (gel kindly supplied by Dr. N. Kalsheker).

to the enzyme Taq 1 appears to be promising. About 5% of a healthy population have a sequence change in the 3' flanking region of the α_1AT gene which results in the loss of a restriction site for the endonuclease Taq 1. If DNA is digested with this enzyme, run on a size exclusion gel and then screened with an appropriate gene probe, 2 fragments can be identified of approximately 1.4 and 0.9 kilobases (kb) in length in most people (95% of the population). In the remainder, loss of the Taq 1 restriction site results in the generation of a single larger DNA fragment of about 2.3 kb (see Fig. 1).

When similar techniques are applied to DNA isolated from unrelated patients with emphysema the incidence of the polymorphism rises to approximately 20%.[25,26]

The nature of this polymorphism and its significance have yet to be defined. It may be a marker for abnormalities in other genes yet to be identified or relate to the α_1AT gene itself. Preliminary studies suggest that basal plasma concentrations of α_1AT are normal.[25] However α_1AT is an acute phase protein and concentrations rise rapidly during periods of inflammation. Indeed α_1AT production can be increased in liver and monocytic cells by a variety of factors including IL6[27], GMCSF[28] and endotoxin.[29] Work is currently under way to determine factors which modulate α_1AT gene function and whether polymorphisms of the gene alter regulation of its synthesis or activity.

The importance of these studies rests not only in the identification of the Taq 1 polymorphism but also the role of cells of monocyte/macrophage lineage. These cells are found in large numbers in the lung and will be present in close proximity to the area where emphysema occurs. The mRNA species produced by these cells is different to that produced by liver cells[30] and is modulated by different factors.[27] The relevance of these observations will be discussed later.

Despite more than 25 years of study the role of α_1AT in the pathogenesis of emphysema, even with severe deficiency, remains far from clear. The message would appear to be that the pathogenesis of emphysema is not simple and other factors may be as important as, or more important than α_1AT.

Other Anti-Elastases

Most of the preceding discussion has been based upon the assumption that α_1AT is the crucial anti-elastase in the lung. However, others have been identified and recent studies have started to investigate their potential role in emphysema.

The most well characterised inhibitor is an 11 kDa protein produced by serous glands and clara cells in the lung,[31] called antileucoprotease (ALP), or more recently secretory leucoprotease inhibitor (SLPI). Several workers have identified a variety of other inhibitors isolated from the lung secretions called BLPI[32], BMPI[33] and BSITE.[34]

Some of these inhibitors have immunological identity,[35] as well as similar inhibitory characteristics,[36] and may in reality represent fragments or isoforms of the same protein.

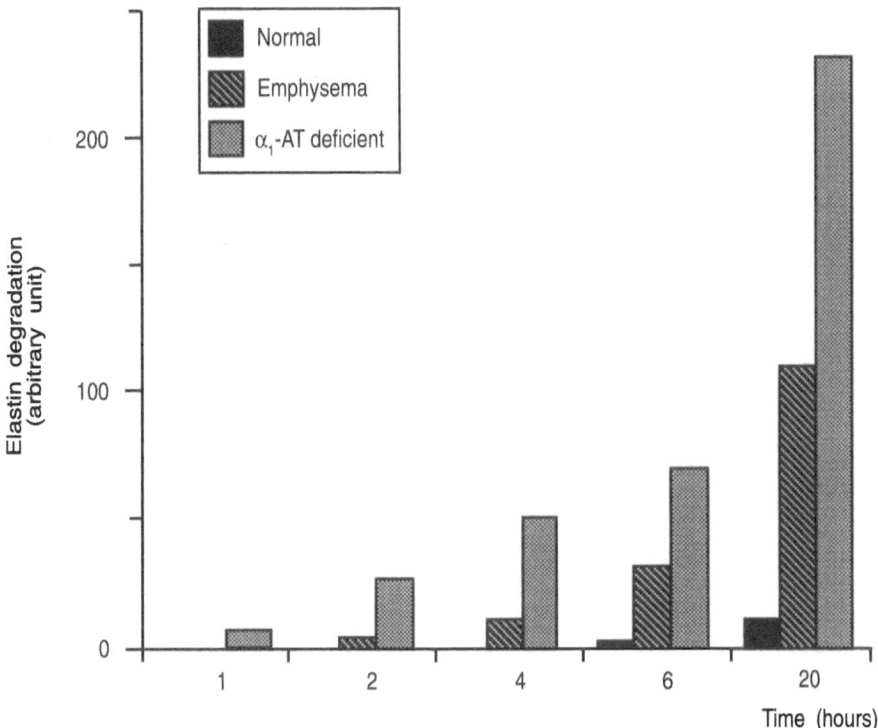

Fig. 2. The effect of mixing NE with excess proteinase inhibitor incubated in the presence of elastin. Fixed amounts of NE, elastin and total moles of inhibitor were used. The inhibitor mixtures were made as indicated to represent α_1AT deficient lavage, emphysema lavage and normal lavage (see text).

Early studies dismissed the role of these inhibitors in the pathogenesis of emphysema because the majority of the anti-elastase function in the lower respiratory tract could be accounted for by α_1AT.[5] These low molecular weight anti-elastases were thought to be a feature of bronchial secretions and thus play a role in the protection of the upper airways. However these inhibitors are more labile than α_1AT as well as being reversible inhibitors of NE.

Subsequent studies suggested that techniques used in early work could inactivate the lung anti-elastases[37] and underestimate their presence in assays based on their function.[37,38]

Thus the anti-elastase screen of the respiratory tract is a mixture of inhibitors with varying concentrations and different affinities for NE. In a series of experiments it

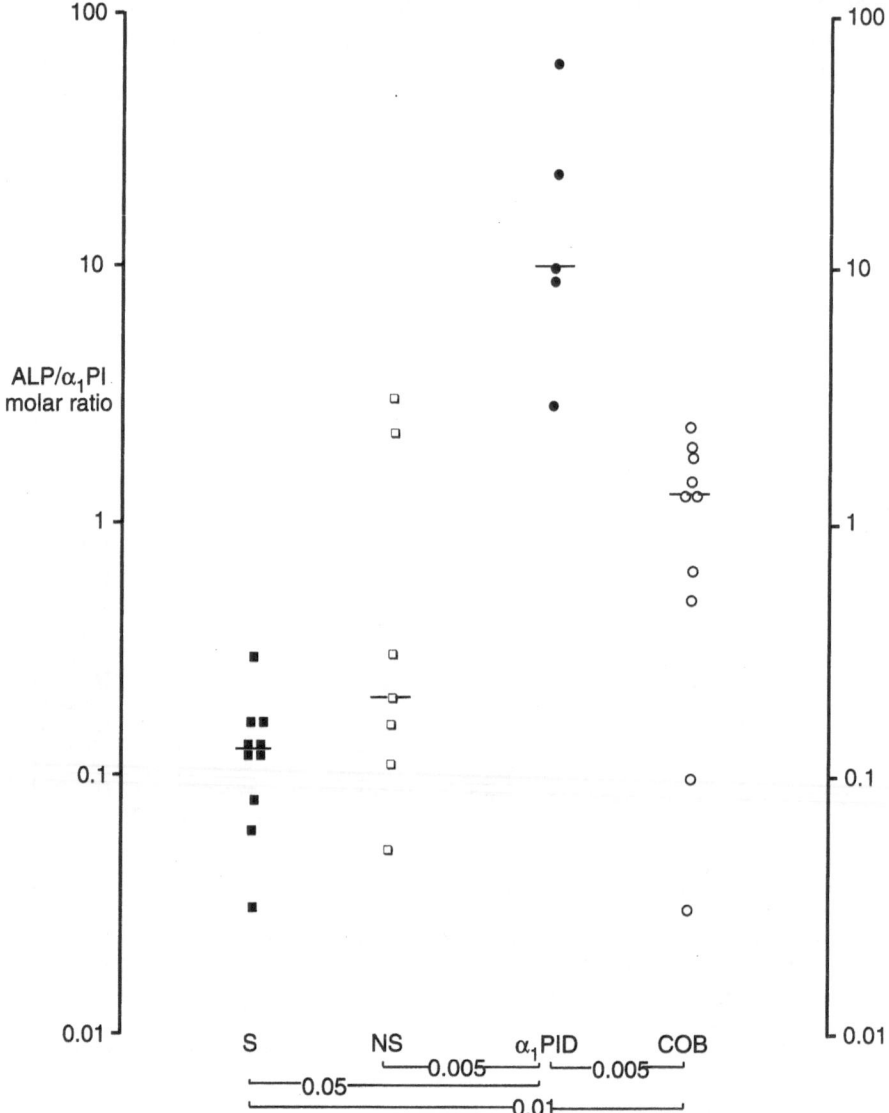

Fig. 3. Molar ratio of ALP to α_1AT in lavage samples obtained from healthy smokers (■) and non-smokers (❑), as well as patients with emphysema who have α_1AT deficiency (α_1PID) (●) or normal α_1AT (COB) (○). The significance of differences between groups is shown. Note 2 healthy non-smokers have high ALP/α_1AT ratios. Reproduced with permission from European Journal of Respiratory Disease.

was shown that the relative concentrations of the inhibitors in biological fluids, as well as artificially prepared mixtures, influence the ability of the fluid to control NE, especially in the presence of elastin.[39] The results, summarised in figure 2, suggest that the greater the ratio of lower affinity inhibitors such as ALP to α_1AT the less the mixture is able to control NE, particularly in the presence of prolonged incubation with elastin.[39] Indeed ratios of ALP to α_1AT vary between subjects. In healthy non-smokers the ratio is about 1:10 and in patients with established emphysema it is almost equimolar. In the presence of α_1AT deficiency the ratio is reversed (10:1). Healthy smokers are similar to non-smokers but some subjects have an ALP/α_1AT ratio similar to subjects with emphysema. It is possible that these findings in emphysema represent effect rather than cause, although it is tempting to speculate that the subset of healthy subjects with high ALP/α_1AT ratios may represent the "susceptible" individuals. These data are summarised in figure 3. This concept of the importance of inappropriate ratio of reversible to non-reversible inhibitors has been supported recently by Stone and colleagues.[40] These workers showed that when hamsters were treated with NE in the presence of an excess of a reversible synthetic inhibitor the resulting emphysema was increased. Clearly the nature and role of inhibitor mixtures in the lung is worthy of further study.

The Cell

Most of the studies of proteinases and antiproteinases have been based on our understanding of the way these proteins interact in solution. Although the neutrophil has been considered to be the major source of NE in the lung, there have been relatively few studies of the activation of this cell and its interaction with connective tissues. Recent work by Burnett et al.[41] has shown that neutrophils from patients with emphysema demonstrate an increased chemotactic response and ability to digest connective tissue. The implication of these findings is that for a given stimulus more cells will be recruited to the lungs of emphysema patients than healthy age matched control subjects. Furthermore once the cells enter the lung it is likely that they possess an increased facility to destroy connective tissue. It remains to be determined whether these results represent the extreme of the normal range and hence the population at risk, or whether these changes are the consequence of established lung disease. Certainly it is possible to enhance connective tissue degradation by neutrophils using endotoxin and tumour necrosis factor[42] both of which may be released in the presence of established lung disease. Further studies will be necessary to clarify these possibilities.

Cell/Substrate Interaction

Once neutrophils enter the lung they are in close contact with interstitial tissue and it is within this region of close contact that the majority of tissue degradation occurs.

56

Campbell and his colleagues have investigated this process of cell/connective tissue interaction in a series of elegant studies which have suggested that this is a privileged site for enzyme/substrate interaction.[43,44] Once the cells have become adherent to the substrate, tissue degradation continues even in the presence of surrounding inhibitors whether of low or high molecular weight. The studies suggest that the distance between the cell membrane (where the enzymes are released) and the connective tissue (where degradation occurs) is too small to admit protein molecules. The implication of these studies is that once cells gain access to the interstitium of the lung, tissue degradation will occur even in the presence of proteinase inhibitor. If this is the case it could be argued that proteinase inhibitors have very little role to play in the pathogenesis of emphysema and that cell recruitment and activation are the critical factor. However, although this may be partially correct, proteinase inhibitors may still play a vital role for 3 reasons.

Firstly the presence of active proteinase inhibitors may limit the tissue degradation to the area of cell/substrate interaction alone. Incidental proteolytic degradation in the area around the cell will be reduced because of inhibition of enzyme by the surroundings inhibitors. However although this limitation can be demonstrated using a 2-dimensional matrix[43] it is difficult to envisage such a sequence of events

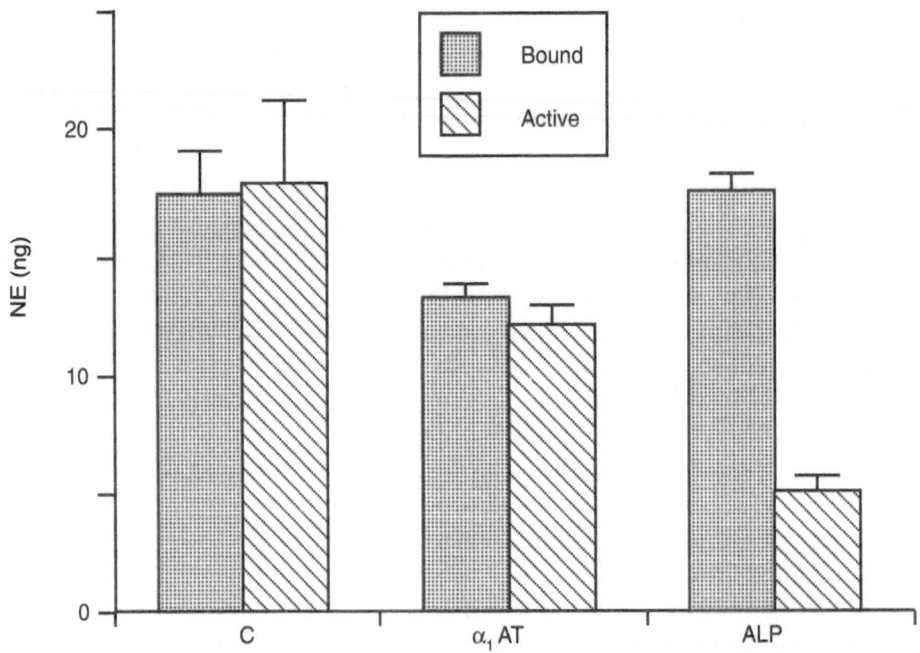

Fig. 4. The amount of NE bound to elastin is shown together with elastolytic activity for control samples and in the presence of α_1AT and ALP. For details see text and reference 47. Results are mean ±SE.

in a 3-dimensional matrix where the cell is surrounded by, and in contact with, connective tissue.

The second alternative is that the inhibitors play a role before or after the cell has passed through the connective tissue. ALP is a cationic protein and will tend to accumulate on connective tissue substrates such as elastin. Indeed immuno-histochemical studies have confirmed the association of ALP with lung elastin.[45] If this inhibitor is bound directly to the elastin (i.e. not via elastase bound to elastin), it may form an *in situ* elastase screen which will protect elastin even when cells adhere and move through or over the connective tissue. In this case proteolysis will only occur in areas where subcellular release of enzymes is greater than the inhibitor present. Alternatively the inhibitors may play a role after the cell has passed by inactivating NE that has become bound to the elastin. Bruch and Bieth demonstrated that ALP had the ability to inactivate elastolysis of elastin previously exposed to NE, whereas α_1AT was relatively ineffective.[46]

More recently these studies have been extended by Morrison et al.[47] to investigate the mechanism of inhibition of NE bound previously to elastin. The evidence confirms that ALP is an effective inhibitor of NE bound to elastin and this is achieved by the ALP interacting with the elastin/NE complex. On the other hand α_1AT inhibits elastin-bound NE poorly but by dissociation of the NE from the elastin. These data are summarised in figure 4 and suggest that a major role for ALP could be the inhibition of NE released by neutrophils, that has already bound to elastin whereas α_1AT will inhibit any NE which remains unbound to substrate.

The final way that proteinase inhibitors may influence connective tissue degradation is by direct effect on cell migration. A variety of studies have suggested that serine proteinases may be involved in activation and migration of cells via membrane receptor associated events. The exact mechanisms are unknown but serine proteinase inhibitors can reduce the chemotactic response of neutrophils to the peptide FMLP.[48] Furthermore, inactivation of α_1AT by oxidation or cleavage of the active site results in the generation of protein that is itself chemotactic.[48] Since cleaved and oxidised α_1AT have been identified in the lungs of patients with emphysema (see above), these inactivated forms may actually increase neutrophil recruitment in some instances. Further studies of this process are under way.

Role of the Monocyte/Macrophage

The neutrophil has received a great deal of attention as the source of NE (the putative enzyme causing emphysema). Recent studies have shown that a subpopulation of circulating monocytes also contain the enzyme[49] and this may account for the ability of these cells to degrade connective tissue. Monocytes are circulating cells which are believed to be the precursors of tissue macrophages. In order to move from the vascular space into the tissues these cells first need to adhere to vascular endothelium. Studies by Owen et al.[50] have shown that only about 20% of circulating monocytes will adhere spontaneously.

Studies are currently under way to determine whether these cells which adhere spontaneously are the same ones that contain NE and are thus capable of damaging lung connective tissue. However monocyte/macrophages can produce α_1AT and therefore also have the ability to protect lung connective tissue from NE, although the degree of α_1AT production also varies amongst monocyte subpopulations.[51]

Finally, recent work has also highlighted the role of the macrophage in the removal of senescent or dying neutrophils.[52] This "clearance" mechanism may be a further way of protecting the lung from enzymes released as the neutrophils die.

Further work is required to clarify the role of monocytes and macrophages in the pathogenesis of emphysema. The role of these cells in the production of other proteinases and inhibitors will be discussed below.

The major part of this review has concentrated on factors which affect the relationship of NE and its inhibitors in emphysema. It supposes that NE is the sole enzyme responsible for the disease. However although NE can produce emphysema in animal models other enzymes present, or potentially present, in the lung are also capable of degrading elastin.

Neutrophil Elastase

NE activity has been shown to be present in sputum from patients with chronic lung infections due to bronchiectasis[53] or cystic fibrosis[54] and in sputum from patients with emphysema, but only during infective exacerbations.[23] In contrast to the large numbers of studies that have used bronchoalveolar lavage (BAL) to characterise the proteinase inhibitors in the lower airways, there have been few attempts to identify the proteinases that are present. NE activity has been reported in BAL from patients with ARDS[55] and acute bacterial pneumonia[56] but was not identified in BAL from healthy subjects including cigarette smokers.[57-59] Similarly, in the single study that has investigated elastases in BAL from subjects with emphysema (in the absence of infection) no evidence was found for the presence of NE activity.[60] The results to date therefore suggest that free NE activity is only present within the airways following a massive influx of neutrophils in response to infection or trauma.

Nevertheless, increased numbers of neutrophils are found in BAL obtained from patients with emphysema.[6] The absence of NE activity suggests that the amount of NE released in the airways is usually insufficient to overwhelm the inhibitors that are present.

Indeed, neutrophils seem reluctant to release significant quantities of their azurophil granule contents, even when treated *in vitro* with stimulating agents.[61-63] We have confirmed that several cytokines, although stimulating extracellular proteolysis by neutrophils, fail to cause a general degranulation and release of NE (Fig. 5). Furthermore, there is evidence to shown that macrophages can recognize and engulf senescent neutrophils, intact, before they have the opportunity to disintegrate and therefore release their potentially harmful enzymes.[52] These features, along with the presence of proteinase inhibitors such as α_1AT, indicate protective mechanisms normally prevail and prevent tissue damage due to uncontrolled extracellular activity of serine proteinases such as NE.

It may be that pathological tissue damage results from an "uncoupling" of the normal mechanisms that control the release of proteinases by neutrophils which, in subjects with emphysema, have an increased potential for recruitment and tissue damage.[41,42] As indicated above, it is possible that NE does mediate tissue damage, but only intermittently, perhaps during infective exacerbations when particularly

Effects of cytokines on proteolysis and elastase release

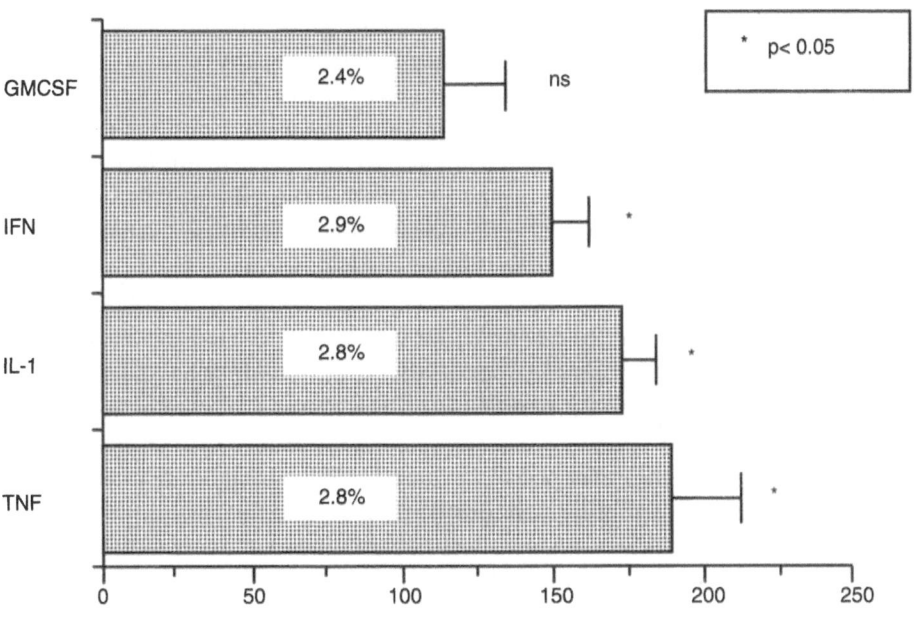

Fibronectin proteolysis % Control

Fig. 5. The effects of 25 Units/ml of Granulocyte-Macrophage Colony-Stimulating Factor (GMCSF), 10^3 Units/ml Interferon-γ (IFN), 25 Units/ml Interleukin -1α (IL-1) and 10^3 Units/ml of Tumour Necrosis Factor-α (TNF) on extracellular proteolysis of radiolabelled fibronectin by neutrophils. Results are shown as mean % (+SEM) of control neutrophils to which no cytokines were added. Note that GMCSF had no significant effect on proteolysis, in contrast to the other cytokines. The numerical values in the bars show the proportion of total cellular neutrophil elastase released by the cells during the culture period. None of the cytokines induced release of more elastase than was seen in control cultures (mean 3.6%).

large numbers of (activated) neutrophils enter the lung. Alternatively, BAL might not represent the sites in the lung where tissue damage occurs. Damiano et al.[64] attempted to probe the interstitial areas for the presence of NE, using immunohistochemical techniques. They showed that NE was present within neutrophils identified in the interstitium of patients with emphysema. Furthermore, NE was also located on interstitial elastic fibres. It has been claimed that this observation was artefactual[65] but the most significant, and perhaps most convincing, evidence from Damiano's work was a correlation between the amount of NE associated with interstitial elastin and the degree of local emphysematous changes. In summary, the evidence from BAL studies does not support the notion that large amounts of free NE activity are present in the lungs of patients with emphysema.

Nevertheless, a compelling hypothesis is emerging to explain how NE might be responsible for elastin destruction in this disease although it will undoubtedly eventually prove to be an oversimplification.

Chemotactic factors produced as a result of inflammatory insults, including cigarette smoke, attract neutrophils to the lung interstitium. These cells are particularly responsive to chemotactic factors.[41] This might occur as a primary difference or result from the release of excess mediators from the inflamed lung. These neutrophils have an increased potential to mediate extracellular connective tissue proteolysis[41,42] caused by limited release of NE in the pericellular space, from which inhibitors such as α_1AT are excluded.[43,44,66] The pericellular elastolysis would be potentiated by the presence of inflammatory mediators (Fig. 5) and although inhibitors are largely excluded from the site, the process might be exacerbated in subjects with α_1AT deficiency. The relatively small amounts of NE released by the cells would remain bound to elastin or be rapidly inactivated by the available inhibitors, explaining the presence of the enzyme on interstitial elastin[64] but an absence of free enzyme activity in samples of BAL. The reasons why some individuals should be susceptible to such mechanisms and therefore develop disease remain to be explained. Clearly the entire process is complex and it is therefore likely that the causes of disease susceptibility will also be far from simple.

Although NE remains a major candidate for the putative mediator in the progression of emphysema, other proteinases which can hydrolyse elastin have been described.

Metalloelastases and Serine Elastases

In addition to NE, neutrophils contain another elastase, proteinase 3. This serine proteinase, which is identical to a protein called myeloblastin and has considerable structural homology with NE[67] is a more effective elastase than NE at pH 6.5 but is less active at neutral pH.[68] Little is known yet about the normal functions, extracellular activity or pathological role of this enzyme although it has been shown to produce experimental emphysema in hamsters.[68]

Although analyses of BAL from healthy or emphysematous patients[57-60] failed to identify specifically the presence of NE activity, elastolytic activity was detected in two of these studies. Niederman et al.[58] reported that elastolytic activity in BAL from healthy smokers was higher than that from non-smokers. The elastolytic activity was largely inhibited by ethylene-diamine tetraacetic acid (EDTA), suggesting metalloproteinase activity, but serine elastase activity was also present. A later study[60] showed higher levels of elastolytic activity in BAL from patients with emphysema than that from healthy subjects. This activity was also shown to be composed of serine and metalloproteinases. The enzymes were large but sensitive to treatment with lipase or detergent, suggesting they were associated with lipid or possibly fragments of cell membranes. The identities and cellular sources of these enzymes remain unknown, as do their potential role in disease processes. It remains possible that some of the serine elastase activity represented membrane-associated NE or proteinase 3, although studies so far indicate that antibodies to NE failed to adsorb the activity from BAL.[60]

The question remains: why was serine elastase activity present in these BAL samples in the absence of free NE activity? It may be that the lipid-associated enzymes were present in amounts greatly exceeding NE and therefore escaped inactivation by endogenous inhibitors. Alternatively, it is possible that the endogenous serine proteinase inhibitors have a low affinity for the lipid-associated enzymes and enzyme-inhibitor complexes were either not formed, or dissociated during the long incubation period with substrate used in the studies.

Dissociation of enzyme from inhibitor might also explain the presence of metalloproteinase activity in BAL despite the presence of Tissue Inhibitor of Metalloproteinases (TIMP), which has been identified in lung secretions.[69] Senior et al.[70] have recently identified the presence of a metalloelastase expressed by human alveolar macrophages.

Despite several previous attempts, no convincing evidence had been presented for a human counterpart to the mouse macrophage elastase[60,71], perhaps because it has now been shown that the human macrophages express a metalloelastase only after 24 hours, incubation on the elastin substrate.[70] The enzyme activity was not demonstrated in cell culture supernatants. Several explanations were offered for this observation.

The enzyme may have been active only in the pericellular space, where it would be protected from inactivation by inhibitors such as TIMP. In addition, the metalloelastase might be membrane-bound and the authors speculated that it could contribute to the enzyme activity seen in BAL samples. Clearly, further work will be required in order to clarify the role of metalloelastases in tissue destruction, but it now appears that macrophages also might contribute to metalloelastase-mediated elastinolysis through pericellular enzyme activity.

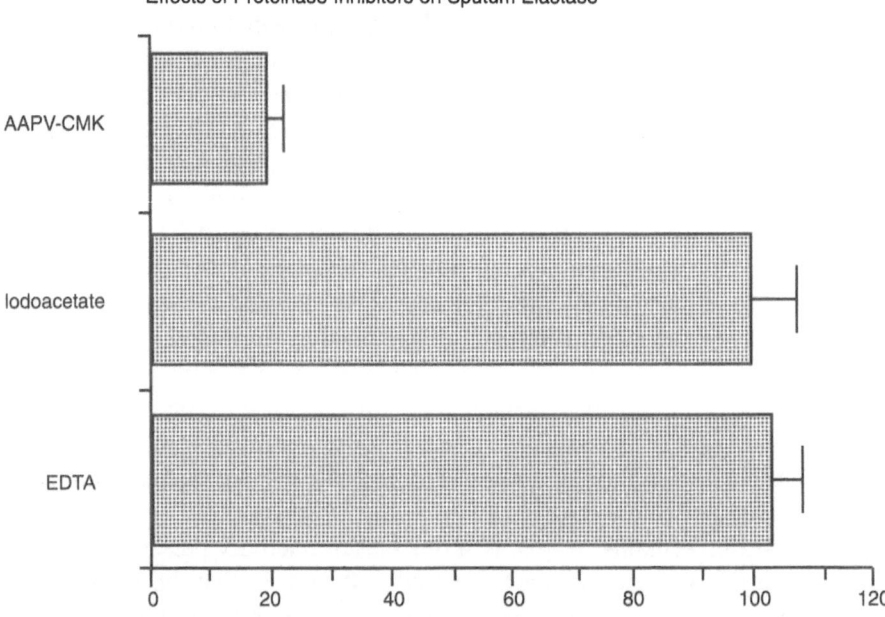

Effects of Proteinase Inhibitors on Sputum Elastase

Elastin proteolysis %Control

Fig. 6. The effects of the serine neutrophil elastase inhibitor. Succ-Ala-Ala-Pro-Val-Chloromethylketone (AAPV-CMK; 1 μmol/l), the cysteine proteinase inhibitor Iodoacetate (1 mmol/l) and the metalloproteinase inhibitor, ethylenediamine tetraacetic acid (EDTA; 1 mmol/l) on elastin proteolysis by sputum samples (n= 5) to which 2 mmol/l dithiothreitol had been added to activate any cysteine proteinases present. The results are shown as mean %(+SEM) of control values, to which no inhibitors were added. The sputum samples were added to fluorescein-labelled elastin and incubated at 37°C, pH 6. The average amount of elastin digested by the control samples, was 0.15 (SEM 0.03) mg elastin/hr/ml sputum. The elastase activity in sputum was significantly reduced by AAPV-CMK but not iodoacetate or EDTA, suggesting that the activity was due to serine neutrophil elastase.

The potential for lung cells other than neutrophils or macrophages to produce serine elastases or metalloelastases is not known. Fibroblasts from other organs have been reported to produce elastases[72] but lung cells have not been studied although they would be well sited to contribute to interstitial proteolysis. Similarly, the possibility that lung epithelial cells might synthesise such enzymes remains speculative.

Many bacteria produce proteolytic enzymes although most do not have elastolytic properties. *Pseudomonas aeruginosa*, however, produces a potent metalloelastase.[73] The role of bacterial elastases in the pathogenesis of emphysema is not clear although they would probably only have the potential to contribute during infective exacerbations.

Cysteine Elastases

Only one human cysteine proteinase, cathepsin L, has been shown definitively to have activity against elastin. The elastolytic activity was equal to that of NE but only at acid pH.[74] Nevertheless, it has been proposed that acid conditions, suitable for the activity of the acid cathepsins, exist in the pericellular area of adherent macrophages.[75] Cathepsin L is a product of human alveolar macrophages[76] and a preliminary report suggested that synthetic inhibitors of this enzyme could prevent elastin proteolysis by these cells.[77]

These results would appear to contradict those of Senior et al.[70] who were able to inhibit elastin proteolysis by macrophages with metalloproteinase inhibitors including TIMP. Since it was shown that the cells had to be incubated for 24 hours on elastin before the metalloelastase activity was expressed, time and the microenvironment might determine the nature of the elastase produced by alveolar macrophages.

Cathepsin L might therefore be another enzyme involved in cell-associated elastin destruction. This enzyme, however, has not been detected in lung secretions although another cysteine proteinase, cathepsin B has been identified in sputum and BAL samples.[78,79] Cathepsin B is also a product of macrophages and it might be expected that the presence of cathepsin B activity would be accompanied by that of cathepsin L. Nevertheless, we have shown that most of the elastolytic activity in sputum samples, incubated with elastin, was inhibited by serine proteinase inhibitors; cysteine proteinases appared to contribute nothing (Fig. 6).

Furthermore all cysteine proteinase activity in lung secretions[78,79] was inhibited by a specific cathepsin B inhibitor, suggesting that there was no cathepsin L activity present. There are several possible explanations for this apparent discrepancy. Firstly, the natural inhibitors of cysteine proteinases, the cystatins and kininogens, most of which have been identified in lung secretions[80], bind to and inhibit cathepsin L more avidly than cathepsin B.[81] It has been observed that a significant proportion of the cathepsin B in sputum is bound to inhibitors but readily dissociates.[82] Cathepsin L would not dissociate from the inhibitors so easily and might remain bound and effectively inhibited.

Secondly, cathepsin L is unstable at alkaline pH.[76] The lysosomal form of cathepsin B is also irreversibly denatured at pH > 7 and thus the presence of cathepsin B activity in lung secretions, where the pH often exceeded 7.0[78] seemed paradoxical. Cathepsin B, however, has been shown to exist in several forms. That of lung secretions, thought to be a truncated form of pro-cathepsin B, is of larger M_r (37K) than the lysosomal enzyme and is stable at alkaline pH.[82] The source of this form of cathepsin B is not known although there is evidence to suggest that it is the product of partial digestion of pro-cathepsin B by NE (D. Buttle, unpublished). Alternatively, cathepsin B might be the secreted product of cells other than macrophages. For instance, bronchial epithelium has been shown to contain the enzyme[83] and might therefore synthesise and secrete the alkaline-stable form. There is no evidence to date to suggest that alkaline-stable forms of cathepsin L exist. If such a form does not exist, the extracellular role of this elastase will definitely be confined to the limited acid environment immediately around cells.The role of cathepsin B as an extracellular enzyme is also not known. It will only digest substrates at acid pH but clearly the lung form can survive the extracellular alkaline conditions in lung secretions. It may therefore have the potential to contribute to connective tissue digestion since some connective tissue proteins including proteoglycan and collagen are substrates for this enzyme.[84] The lysosomal form of cathepsin B was reported not to be elastolytic[84] but a recent presentation[85] suggested that bovine cathepsin B was not only able to digest elastin, but could also produce emphysema when instilled into the trachea of hamsters. This preparation of enzyme was from a commercial source and the purity was not assessed. Nevertheless, these results, and others, suggest further studies are indicated into the role of cysteine proteinases in the pathogenesis of human emphysema.

Conclusion

This review has covered several aspects of proteinases and their inhibitors in the pathogenesis of emphysema. As such it has concentrated upon elastase and anti-elastases, assuming that the critical target in the generation of emphysema is lung elastin. Most of the evidence for the central importance of elastin is circumstantial, although animal experiments in which elastin synthesis has been modified suggest that elastin degradation is of some importance. However neutrophil elastase is not specific for elastin and will damage many other components of the extracellular matrix. It remains possible that enzymes with other specificities and hence other inhibitors may also play a role in the pathogenesis of emphysema. Nevertheless the information and concepts generated to date provide a good template for future studies.

References

1. Eriksson S.: Studies in α_1 antitrypsin deficiency in the lower respiratory tract of humans. Acta Med. Scand. 1965; 177 (suppl): 1-85
2. Gross P., Pfitzer E.A., Tolker E., Babyok M.A:, Kaschok M.: Experimental emphysema. Its production with papain in normal and silicotic rats. Arch. Envir. Health. 1964; 11: 50-58
3. Beatty K., Bieth J.G., Travis J.: Kinetics of association of serine proteinase inhibitor and with α_1 antichymotrypsin. J. Biol. Chem. 1980; 255: 3931-3934
4. Senior R.M., Tegner H., Kuhn C., Ohlsson K., Starcher B.C., Pierce J.A.: The induction of pulmonary emphysema with human leucocyte elastase. Am. Rev. Respir. Dis., 1977; 116: 469-475

5. Gadek J.E., Fells G.A., Zimmerman R.L., Rennard S.I., Crystal R.G.: Antielastases of the human alveolar structures. Implications for the protease-antiprotease theory of emphysema. J. Clin. Invest., 1981; 68: 889-898

6. Morrison H.M., Kramps J.A., Burnett D., Stockley R.A.: Lung lavage from patients with alpha-1-proteinase inhibitor deficiency or chronic obstructive bronchitis: antielastase function and cell profile. Clin. Sci. 1987; 72: 373-381

7. Larsson C.: Natural history and life expectancy in severe α_1-antitrypsin deficiency PiZ. Acta Med. Scand., 1978; 204: 345-351

8. Janus E.D., Philips N.T., Carrell R.W.: Smoking, lung function and α_1 antitrypsin deficiency. Lancet, 1985; i: 152-154

9. Silverman E.K., Pierce J.A., Province M.A., Rao D.C., Campbell E.J.: Variability of pulmonary function in α_1-antitrypsin deficiency: clinical correlates. Ann. Intern. Med. 1989; 111: 982-991

10. Cohen B.H., Ball W.C., Bias W.B., Brashears S., Chase G.A., Diamond E.L., Hsu S.H., Kreiss P., Levy D.A., Menkes H.A., Permutt S., Tockman M.S.: A genetic-epidemiologic study of chronic obstructive pulmonary disease. Johns Hopkins Med. J. 1975; 137: 95-104

11. Ward P.A., Talamo R.C.: Deficiency of the chemotactic factor inactivators in human serum with α_1-antitrypsin deficiency. J. Clin. Invest., 1973; 52: 512-519

12. Mierzwa S., Chan S.L.: Chemical modification of human α_1-proteinase inhibitor by tetranitromethane. Structure-function relationship. Biochem. J. 1987; 246: 37-42

13. Gadek J.E., Fells G.A., Crystal R.G.: Cigarette smoking induces functional antiprotease deficiency in the lower respiratory tract of humans. Science, 1979; 206: 1315-1316

14. Carp H., Miller F., Hoidal J.R., Janoff A.: Potential mechanism of emphysema: α_1-proteinase inhibitor recovered from the lungs of cigarette smokers contains oxidised methionine and has decreased elastase inhibitory capacity. Proc. Natl. Acad. Sci. USA., 1982; 79: 2041-2045

15. Abboud R.T., Fera T., Richter A., Tabona M.Z., Johal S.: Acute effect of smoking on the functional activity of α_1-protease inhibitor in bronchoalveolar lavage fluid. Am. Rev. Respir. Dis. 1985; 131: 79-85

16. Stone P.J., Calore J.D., McGowan S.E., Bernado J., Snider G.L., Franzblau C.: Functional α_1-proteinase inhibitor in the lower respiratory tract of smokers is not reduced. Science, 1983; 221: 1187-1189

17. Boudier C., Pelletier A., Pauli G., Bieth J.G.: The functional activity of α_1-proteinase inhibitor in bronchoalveolar lavage fluids from healthy human smokers and non-smokers. Clin. Chim. Acta, 1983; 132: 309-315

18. Afford S.C., Burnett D., Campbell E.J., Cury J.D., Stockley R.A.: The assessment of α_1 proteinase inhibitor form and function in lung lavage fluid from healthy subjects. Biol. Chem. Hoppe Seyler., 1988; 369: 1065-1074

19. Stockley R.A., Afford S.C.: Qualitative studies of lung lavage α_1-proteinase inhibitor. Hoppe Seylers Z. Physiol. Chem., 1984; 365: 503-510

20. Campbell E.J., Endicott S.K., Rios-Mollineda R.A.: Assessment of oxidation of α_1-proteinase inhibitor in bronchoalveolar lining fluid by monoclonal immunoassay. Comparison of smokers and non-smokers. Am. Rev. Respir. Dis. 1987; 135, n° 4, part 2; A156

21. Morrison H.M., Burnett D., Stockley R.A.: The effect of catalase and methionine-S-oxide reductase on oxidised α_1-proteinase inhibitor. Biol. Chem. Hoppe Seyler. 1986; 364: 371-378

22. Abrams W.R., Weinbaum G., Weissbach L., Weissbach H., Brot N.: Enzymatic reduction of oxidised α_1-proteinase inhibitor restores biological activity. Proc. Natl. Acad. Sci. USA., 1981; 78: 7483-7486

23. Stockley R.A., Burnett D.: α_1-antitrypsin in infected and non-infected sputum. Am. Rev. Respir. Dis., 1979; 120: 1081-1086

24. Jochum M., Pelletier A., Boudier C., Pauli G., Bieth J.G.: The concentration of leukocyte elastase-α_1-proteinase inhibitor complex in bronchoalveolar lavage fluids from healthy subjects. Am. Rev. Respir. Dis. 1985; 132: 913-914

25. Kalsheker N.A., Hodgson I.J., Watkins G.L., White J.P., Morrison H.M., Stockley R.A.: Deoxyribonucleic acid (DNA) polymorphism of the $\alpha1$ antitrypsin gene in chronic lung disease. Br. Med. J., 1987; 294: 1511-1514

26. Kalsheker N.A., Watkins G.L., Hill S., Morgan K., Stockley R.A., Fick R.B.: Independent mutations in the flanking sequence of the α_1-antitrypsin gene are associated with chronic obstructive airways disease. Dis. Markers., 1990; 8: 151-57

27. Perlmutter D.H., May L.T., Sehgal P.B.: Interferon $\beta2$/Interleukin 6 modulates synthesis of α_1-antitrypsin in human mononuclear phagocytes and in human hepatoma cells. J. Clin. Invest., 1989; 84: 138-144

28. Owen C.A., Todd H.C., Wilson R., Stockley R.A.: Differential regulation of α_1-antitrypsin production by adherent and non adherent monocytes. Am. Rev. Resp. Dis. 1990; 141 (N° 4, part 2): A113

29. Barbey-Morel C., Pierce J.A., Campbell E.J., Perlmutter D.H.: Lipopolysaccharide modulates the expression of $\alpha1$-proteinase inhibitor and other serine proteinase inhibitors in human monocytes and macrophages. J. Exp. Med. 1987; 166: 1041-1054

30. Perlino E., Cortese R., Ciliberto G.: The human alpha-1-antitrypsin gene is transcribed from two different promoters in macrophages and hepatocytes. EMBO J., 1987; 6:2767-2772

31. Mooren H.W.D., Kramps J.A., Franken C., Meijer C.J.L.M., Dijkman J.H.: Localisation of a low molecular weight bronchial protease inhibitor in the human peripheral lung. Thorax, 1983; 38: 180-183

32. Smith C.E., Johnson D.A.: Human bronchial leucocyte proteinase inhibitor. Rapid isolation and kinetic analysis with human leucocyte proteinases. Biochem. J., 1985; 225: 463-472

33. Smith S.F., Guz A., Burton G.H., Cooke N.T., Tetley T.D.: Acid-stable low molecular mass proteinase inhibitors in human lung lavage. Biol. Chem. Hoppe Seyler; 1986; 367: 183-189

34. Hochstrasser K., Albrecht G.J., Schonberger O.L., Rasche B., Lemport K.: An elastase-specific inhibitor from human bronchial mucus. Isolation and characterization. Hoppe. Seylers Z. Physiol. Chem., 1981; 362: 1369-1375

35. Keuppers F., Bromke B.J.: Protease inhibitors of tracheo-bronchial secretions. J. Lab. Clin. Med., 1983; 101: 747-757

36. Morrison H.M.: The proteinase-antiproteinase theory of emphysema: time for a reappraisal? Clin. Sci. 1987; 72: 151-158

37. Afford S.C., Stockley R.A., Kramps J.A., Dijkman J.H., Burnett D.: Concentration of bronchoalveolar lavage fluid by ultrafiltration: evidence of differential protein loss and functional inactivation of proteinase inhibitors. Analyt. Biochem. 1985; 151: 125-130

38. Kramps J.A., Morrison H.M., Burnett D., Dijkman J.H., Stockley R.A.: Determination of elastase inhibitory activity of α_1-proteinase inhibitor and anti leukoprotease: different results using insoluble elastin or synthetic low molecular weight substrates. Scand. J. Clin. Lab. Invest., 1987; 47: 405-410

39. Stockley R.A., Morrison H.M.: Elastase inhibitors of the respiratory tract. Eur J. Resp. Dis. 1990; 3(suppl.9); 9-15

40. Stone P.J., Lucey E.C., Snider G.L.: Induction and exacerbation of emphysema in hamsters with human neutrophil elastase inactivated reversibly by a peptide boronic acid. Am. Rev. Resp. Dis. 1990; 141: 47-52

41. Burnett D., Chamba A., Hill S.L., Stockley R.A.: Neutrophils from subjects with chronic obstructive lung disease show enhanced chemotaxis and extracellular proteolysis. Lancet, 1987; ii: 1043-1046

42. Burnett D., Chamba A., Hill S.L., Stockley R.A.: Effect of plasma, tumour necrosis factor, endotoxin and dexamethasone on extra-cellular proteolysis by neutrophils from healthy subjects and patients with emphysema. Clin. Sci., 1989; 77: 35-41

43. Campbell E.J., Senior R.M., McDonald J.A., Cox D.L.: Proteolysis by neutrophils. Relative importance of cell-substrate contact and oxidative inactivation of proteinase inhibitors in vitro. J. Clin. Invest., 1982; 70: 845-852

44. Campbell E.J., Campbell M.A.: Pericellular proteolysis by neutrophils in the presence of proteinase inhibitors: Effects of substrate opsonisation. J. Cell. Biol. 1988; 106: 667-676

45. Willems L.N.A., Otto-Verberne C.J.M., Kramps J.A., ten Have-Opbrock A.A.W., Dijkman J.H.: Detection of anti-leukoprotease in connective tissue of the lung. Histochemistry, 1986; 86: 165-168

46. Bruch M., Bieth J.G.: Influence of elastin on the inhibition of leucocyte elastase by α1-proteinase inhibitor and bronchial inhibitor. Potent inhibition of elastin-bound elastase by bronchial inhibitor. Biochem. J., 1986; 238: 269-273

47. Morrison H.M., Welgus H.G., Stockley R.A., Burnett D., Campbell E.J.: Inhibition of human leukocyte elastase bound to elastin: relative ineffectiveness and two mechanisms of inhibitory activity. Am. J. Respir. Cell Mol. Biol. 1990; 2: 263-269

48. Stockley R.A., Shaw J., Afford S.C., Morrison H.M., Burnett D.: Effect of α_1-proteinase inhibitor on neutrophil chemotaxis. Am. J. Respir. Cell Mol. Biol., 1990; 2: 163-170

49. Campbell E.J., Silverman E.K., Campbell M.A.: Elastase and Cathepsin G of human monocytes. Quantification of cellular content, release in response to stimuli and heterogeneity in elastase-mediated proteolytic activity. J. Immunol., 1989; 143: 2961-2968

50. Owen C.A., Afford S.C., Hill S.L., Burnett D., Stockley R.A.: Monocyte adherence in bronchiectasis and the effect of bacterial lipopolysaccharide. Am. Rev. Respir. Dis. 1989; 139; n° 4, part 2:A127

51. Owen C.A., Afford S.C., Burnett D., Stockley R.A.: Differential α_1-PI synthesis by monocyte subpopulations. Am. Rev. Respir. Dis., 1989; 139: n° 4, part 2:A201

52. Saville J.S., Wyllie A.H., Henson J.E., Walport M.J., Henson P.M., Haslett C.: Macrophages phagocytosis of aging neutrophils in inflammation. Programmed cell death in the neutrophil leads to its recognition by macrophages. J. Clin. Invest., 1989; 83: 865-875

53. Stockley R.A., Hill S.L., Morrison H.M., Starkie C.M.: Elastolytic activity of sputum and its relationship to purulence and to lung function in patients with bronchiectasis. Thorax, 1984; 39: 408-413

54. Goldstein W., Doring G.: Lysosomal enzymes from polymorphonuclear leukocytes and proteinase inhibitors in patients with cystic fibrosis. Am. Rev. Respir. Dis., 1986; 134: 49-56

55. Lee C.T., Fein A.M., Lippmann M.L., Holtzman H., Kimbel P., Weinbaum G.: Elastolytic activity in pulmonary lavage fluid from patients with adult respiratory distress syndrome. N. Engl. J. Med., 1981; 304: 192-196

56. Abrams W.R., Fein A.M., Kucich U., Kueppers F., Yamada H., Kuzmowycz T., Morgan L., Lippmann M., Goldberg S.K., Weinbaum G.: Proteinase inhibitory function in inflammatory lung disease. 1. Acute bacterial pneumonia. Am. Rev. Respir. Dis., 1984; 129: 735-741

57. Janoff A., Raju L., Dearing R.: Levels of elastase activity in bronchoalveolar lavage fluids of healthy smokers and non-smokers. Am. Rev. Respir. Dis., 1983; 127: 540-544

58. Niederman M.S., Fritts L.I., Merrill W.W., Fick R.B., Matthay R.A., Reynolds H.Y., Gee J.B.L.: Demonstration of a free elastolytic metalloenzyme in human lung lavage fluid and its relationship to α_1-antiprotease. Am. Rev. Respir. Dis. 1984; 129: 943-947

59. Fera T., Abboud R.T., Richter A., Johal S.S.: Acute effect of smoking on elastase-like esterase activity and immunologic neutrophil elastase levels in bronchoalveolar lavage fluid. Am. Rev. Respir. Dis., 1986; 133: 568-573

60. Burnett D., Afford S.C., Campbell E.J., Rios-Mollineda R.A., Buttle D.J., Stockley R.A.: Evidence for lipid-associated serine proteases and metalloproteases in human bronchoalveolar lavage fluid. Clin. Sci., 1988; 75: 601-607

61. Ozaki Y., Ohashi T., Kume S.: Potentiation of neutrophil function by recombinant DNA-produced interleukin 1α. J. Leuk. Biol. 1987; 42: 621-627

62. Kapp A., Zeck-Kapp G., Danner M., Luger T.A.: Human granulocyte-macrophage colony-stimulating factor: an effective direct activator of human polymorphonuclear neutrophilic granulocytes. J. Invest. Dermatol., 1988; 91: 49-55

63. Yonemaru M., Stephens K.E., Ishizaka A., Zheng H., Hogue R.S., Crowley J.J., Hatherill J.R., Raffin T.A.: Effects of tumor necrosis factor on PMN chemotaxis, chemoluminescence and elastase activity. J. Lab. Clin. Med., 1989; 114: 674-681

64. Damiano V.V., Tsang A., Kucich U., Abrams W.R., Rosenbloom J., Kimbel P., Fallahnejad M., Weinbaum G.: Immunolocalization of elastase in human emphysematous lungs. J. Clin. Invest. 1986; 78: 482-493

65. Fox B., Bull T.B., Guz A., Harris E., Tetley T.D.: Is neutrophil elastase associated with elastic tissue in emphysema? J. Clin. Pathol., 1988; 41: 435-440

66. Schalkwijk J., van den Berg W.B., van de Putte L.B.A., Joosten L.A.B.: Elastase secreted by polymorphonuclear leukocytes causes chondrocyte damage and matrix degradation in intact articular cartilage: escape from inactivation by α1-proteinase inhibitor. Brit. J. Exp. Pathol. 1987; 68: 81-88

67. Jenne D.E., Tschopp J., Ludemann J., Utecht B., Gross W.L.: Wegener's autoantigen decoded. Nature (Lond)., 1990; 346:520

68. Kao R.C., Wehner N.G., Skubitz K.M., Gray B.H., Hoidal J.R.: Proteinase 3. A distinct human polymorphonuclear leukocyte proteinase that produces emphysema in hamsters. J. Clin. Invest., 1988; 82: 1963-1973

69. Burnett D.; Reynolds J.J., Ward R.V., Afford S.C., Stockley R.A.: Tissue inhibitor of metalloproteinases (TIMP) and collagenase inhibitory activity in lung secretions from patients with chronic obstructive bronchitis: the effect of corticosteroid therapy. Thorax, 1986; 41: 740-745

70. Senior R.M., Connolly N.L., Cury J.D., Welgus H.G., Campbell E.J.: Elastin degradation by human alveolar macrophages. A prominent role of metalloproteinase activity. Am. Rev. Respir. Dis., 1989; 139: 1251-1256

71. Bieth J.G.: Elastases: catalytic and biological properties. In: Mecham R.P., (ed.) *Regulation of matrix accumulation.* New York, Academic Press. 1986; 217-320

72. Godeau G., Frances C., Hornebeck W., Brechemier D., Robert L.: Isolation and partial characterization of an elastase-type protease in human vulva fibroblasts: its possible involvement in vulvar elastic tissue destruction of patients with lichen sclerosus et atrophicus. J. Invest. Dermatol., 1982; 78: 270-275

73. Morihara K., Tsuzuki H., Oda K.: Protease and elastase of *Pseudomonas aeruginosa*: inactivation of human plasma α-proteinase inhibitor. Infect. Immun., 1979; 24: 188-193

74. Mason R.W., Johnson D.A., Barrett A.J., Chapman H.A.: Elastinolytic activity of human cathepsin L. Biochem. J., 1986; 233: 925-927

75. Silver I.A., Murrills R.J., Etherington D.J.: Microelectrode studies on the acid environment beneath adherent macrophages and osteoclasts. Exp. Cell. Res. 1988; 175: 266-276

76. Reilly J.J., Mason R.W., Chen P., Joseph L.J., Sukhatme V.P., Yee R., Chapman H.A.: Synthesis and processing of cathepsin L, an elastase, by human alveolar macrophages. Biochem. J., 1989; 257: 493-498

77. Chapman H.A., Reilly J.J., Yee R., Mason R.B.: Synthesis and expression of an elastolytic

enzyme, cathepsin L, by human alveolar macrophages. Am. Rev. Respir. Dis. 1987; 135: 293A

78. Burnett D., Crocker J., Stockley R.A.: Cathepsin B-like cysteine proteinase activity in sputum and immunohistologic identification of cathepsin B in alveolar macrophages. Am. Rev. Respir. Dis., 1983; 128: 915-919

79. Burnett D., Stockley R.A.: Cathepsin B-like cysteine proteinase activity in sputum and bronchoalveolar lavage samples: relationship to inflammatory cells and effects of corticosteroid and antibiotic treatment. Clin. Sci. 1985; 68: 469-474

80. Buttle D.J., Burnett D., Abrahamson M.: Levels of neutrophil elastase and cathepsin B activities, and cystatins in human sputum: relationship to inflammation. Scand. J. Clin. Lab. Invest., 1990; 50: 509-516

81. Barrett A.J., Rawlings N.D., Davies M.E., Machleidt W., Salvesen G., Turk V.: Cysteine proteinase inhibitors of the cystatin superfamily. In: Barrett A.J., Salvesen G. (eds.) *Research monographs in cell and tissue physiology.* Vol. 12; *Proteinase Inhibitors.* Amsterdam, Elsevier, 1986; 515-569

82. Buttle D.J., Bonner B.C., Burnett D., Barrett A.J.: A catalytically active high Mr form human Cathepsin B from sputum. Biochem., J., 1988; 254: 693-699

83. Howie A.J., Burnett D., Crocker J.: The distribution of cathepsin B in human tissues. J. Pathol. 1985; 145; 307-314

84. Barrett A.J.: Human cathepsin B1. Purification and some properties of the enzyme. Biochem. J., 1973; 131: 809-822

85. Lesser M., Padilla M., Cardozo C., Orlowski M.: The intratracheal instillation of cathepsin B causes emphysema in hamsters. Am. Rev. Respir. Dis. 1990; 141: A113

6. Multiple Functions of Neutrophil Proteinases and their Inhibitor Complexes

J. Travis[1], J. Potempa[2], N. Bangalore[1], A. Kurdowska[2]
1. Department of Biochemistry, University of Georgia, Athens, Georgia, USA
2. Institute for Molecular Biology, Jagiellonian University, Krakow, Poland

Introduction

During inflammatory episodes neutrophils are recruited to the site of injury where they function as scavengers by killing bacteria and ingesting and degrading foreign and damaged human proteins. As a consequence of these processes significant quantities of neutrophil proteins are released extracellularly, both through cell leakage and cell death. This places a heavy burden on normal, healthy tissues which may now become susceptible to attack by neutrophil-derived oxidizing agents (myeloperoxidase-derived) and proteinases [primarily elastase (HNE) and cathepsin G (cat G)]. It is not clear as to how the body regulates the activity of myeloperoxidase, although it is likely that this involves the use of catalase in order to reduce H_2O_2 levels; however, to offset the possibility of proteinase damage the body offers a series of inhibitors, primarily plasma derived, which function to specifically inactivate these enzymes. In particular, it is now known that human plasma α_1 proteinase inhibitor (α_1PI) regulates the activity of HNE while cat G is controlled by α_1 antichymotrypsin (α_1Achy).[2]

Several years ago it was shown that decreased levels of α_1PI, occurring because of a genetic abnormality in its secretory rate[3], could be correlated with the development of familial emphysema.[4] This led to the conclusion that HNE was the enzyme responsible for abnormal connective tissue degradation associated with the development of the disease. Since that time enormous efforts have been made to develop elastase inhibitors which could be utilized to supplement the protection given by endogenous α_1PI. However, little attention has been paid to the possibility that tight regulation of HNE (and probably cat G) could affect not only the ability of such enzymes to participate in phagocytosis but also in other lesser understood functions.

Chemical Properties of Human Neutrophil Elastase and Cathepsin G

HNE and cat G are basic proteins bound to acidic matrices which make up the azurophil granule of the neutrophil. The binding of the enzymes is presumably through ionic interactions; however, it is not yet known exactly how degranulation occurs to release each enzyme from the cell. Significantly, the concentration of each of these proteinases inside the cell is exceedingly high, ranging from 10 to 30 micromolar. Indeed, the rate of normal turnover of neutrophils is such that approximately 200 mg of each enzyme are excreted per day. Thus, control of the activities of each of these enzymes must be tightly regulated to ensure the integrity of healthy connective tissue proteins.

HNE is secreted as a mixture of glycoproteins of Mr between 25 and 28 kD. The enzymes differ only in carbohydrate content[6], and this is confirmed by the fact that only a single gene has been found which codes for the synthesis of this protein.[7] Interestingly, the protein is apparently synthesized as a zymogen which is activated by removal of a dipeptide from the amino terminal. However, processing at the carboxy end of the proteins must also occur since an extension of 18 amino acid residues over that found in the mature protein must be synthesized according to the gene sequence.

The enzyme has been crystallized as a complex with the third domain of turkey ovomucoid[8], and its structure determined. As expected, marked homology with the three-dimensional structure of porcine pancreatic elastase was found. However, a major difference was noted in that two carbohydrate moieties were found which could be assigned to areas remote from the active site of HNE. The possible role of these residues will be discussed later.

The properties of cat G are similar to those found for HNE. The protein is also synthesized as a mixture of glycoproteins containing both the amino-terminal and carboxy-terminal extensions.[9] However, its crystal structure has not yet been determined. Significantly, this protein contains nearly 20% arginine and is considerably more basic in nature than HNE.

Chemical Properties of α-1-Proteinase Inhibitor and α-1-Antichymotrypsin

Both α_1PI and α_1Achy are glycoproteins which are synthesized in the liver and transported into the circulation. The molar concentration of α_1PI in plasma is far higher than that of α_1Achy[1], and this is somewhat unusual considering the fact that the enzymes which each appears to regulate are present in approximately equal concentrations inside the neutrophil.

Nevertheless, under normal circumstances these levels seem to be appropriate for proteinase regulation, and it may be that the particular properties of each of the two inhibitors accounts for this apparent anomaly.

Human α_1PI is the most studied of the two inhibitors. It is a single chain glycoprotein of Mr near 52 kD which functions to bind serine proteinases through a 1:1 stoichiometric interaction. This indicates that the protein has a single reactive site which has been identified as involving a P-methionine residue.[10] Significantly, oxidation of this residue by either chemical (smoke) or enzymatic (myeloperoxidase) means markedly reduces its ability to interact with HNE so that the modified inhibitor cannot compete with elastin for binding of this enzyme.[11]

As a result HNE can degrade this substrate in the presence of oxidized inhibitor, suggesting a likely mechanism by which uncontrolled proteolysis may occur in tissues where there appears to be abundant inhibitor present.[12] Since familial emphysema is believed to occur as a result of a genetic deficiency in α_1PI levels in blood and other tissues, resulting in high levels of free HNE in tissues, it is likely that in normal individuals the oxidation of this inhibitor causes the evolution of a "pseudo-genetic" defect, again resulting in free enzyme and concomitant destruction of normal, healthy, connective tissue. Human α_1Achy is also a single chain glycoprotein, having a Mr near 58 kD. This inhibitor functions to regulate the activity of neutrophil cat G, again through formation of a 1:1 complex.[1] Inhibition occurs at a leucyl-seryl reactive site peptide bond.[13]

Unlike α_1PI, there does not seem to be any built-in control mechanism for regulating inhibitor activity, nor are there cases of severe α_1Achy deficiency which result in major, debilitating diseases. Indeed, it is likely that such a deficiency state would be lethal to the fetus.

Both α_1PI and α_1Achy belong to the superfamily of proteins referred to as Serpins.[14] The majority of the members of this family are proteinase inhibitors which apparently contain a loop of amino acid residues exposed on the protein surface and which contain the reactive site peptide bond. While no native Serpin has as yet been crystallized, the structure of a post-complex form of α_1PI (α_1PI*) has been determined.[8]

The data obtained indicates that cleavage of the inhibitor at the reactive site methionyl-seryl peptide bond results in a major conformational change which causes the separation of these two residues by more than 70 angstroms. Clearly, the native protein must have a stressed reactive site loop which, when cleaved, causes irreversible changes in the structure of the now modified, inactive protein. Several examples of such inactivating cleavages are now known,[15,16] indicating that attack at Serpin reactive site loops by proteinases not complexed by a given inhibitor may be a mechanism for decreasing the levels of inhibitory activity in tissues. Thus, chemical oxidation (α_1PI) and/or enzymatic cleavage (α_1PI and α_1Achy, as well as several other Serpins) may be utilized to regulate inhibitory activity.

Functions of Neutrophil Proteinases

One of the major functions of neutrophil proteinases involves the degradation of foreign or damaged proteins during the process of phagocytosis. However, recent work from this laboratory would suggest that such enzymes may have alternate functions. These are described in subsequent sections.

Protein Degradation

With regard to the intracellular degradation of proteins it seems clear that HNE is the more important enzyme. It has previously been shown that it can degrade elastin, collagen, and proteoglycan, although elastin degradation is considerably slower than that found with other elastinolytic enzymes (Table I). In contrast, cat G degradation of elastin is barely detectable, although collagen digestion occurs at nearly the same rate as that shown with HNE. At least one report has suggested that cat G can augment the degradation of elastin by HNE.[17] However, others have not been able to repeat such results.[18] Surprisingly, little has been done to follow up the original observations which indicated that cat G could digest proteoglycan at a faster rate than HNE.[19]

Table I. Comparative elastinolytic activity of mammalian and bacterial proteinases

Enzyme	Activity [a]	Enzyme Class
Porcine Pancreatic Elastase	100	Serine
Dog Pancreatic Elastase	199	Serine
Human Neutrophil Elastase	17	Serine
Human Neutrophil Cathepsin G	3	Serine
Horse Neutrophil Elastase	80	Serine
Human Cathepsin L	50	Cysteine
Thermolysin	419	Metallo
Pseudomonas aeruginosa Elastase	41	Metallo
Streptomyces fradiae Elastase	818	Serine
Flavobacterium immotum Elastase	1190	Metallo

[a] Elastinolytic activity was expressed as % of porcine pancreatic elastase activity.

Bactericidal Activity of Cathepsin G

Several years ago it was reported that cat G could act as an efficient bactericidal agent in the killing of both gram-positive and gram-negative bacteria.[20] This activity was not destroyed by inhibition of the proteinase activity of this enzyme, indicating that a distinct domain existed in cat G which was responsible for this activity. Recently[21], two domains have been found which are probably responsible for killing. Each has been identified by sequence analysis and confirmation of their function has been made through testing of synthetic peptides containing these sequences. The two sequences found were:

a. IVGGR

b. HPQYNQR.

The former is placed as the amino-terminal sequence of cat G and is probably not available in the native protein as a bactericidal agent. However, the latter is likely to be placed on the "backside" of the protein where it is exposed and capable of intercalating with bacterial cell walls. Modification of this peptide has confirmed the importance of both the histidine and tyrosine residues for activity. We are currently evaluating variations in the structure of this peptide to determine whether it might prove useful as a peptide antibiotic.

Role of Neutrophil Proteinases in Cytokine Synthesis

One of the most poorly understood areas of research on inflammation involves the determination of signalling mechanisms for acute phase protein synthesis. While it is believed that this involves the utilization of the cytokine IL-6[22], it is not clear as to how this second message is produced. In this laboratory[23] we have found that complexes of α_1Achy and cat G, as well as proteolytically inactivated α_1Achy (α_1Achy*), can cause IL-6 synthesis in a fibroblast system (Table II). When conditioned medium from cells exposed to these proteins is now incubated with a human HepG 2 cell line there is increased synthesis of the common acute phase proteins, including haptoglobin, fibrinogen, ceruloplasmin, α_1Achy, and α_1PI (Table III). It should be noted that the α_1Achy* tested in this system was formed

Table II. Stimulation of IL-6 by lung fibroblasts stimulated with α-1-antichymotrypsin/cathepsin G complexes

Sample	Stimulation
Control	1.0
Cat G	1.2
α_1Achy	3.1
α_1Achy*	3.6
Phorbol Myristyl Acetate	3.5
α_1Achy: cat G	4.6

The activity shown by native α-1-antichymotrypsin is due to cleavage by proteinases released from fibroblasts during the incubation period.

Table III. Stimulation of synthesis of acute phase proteins in HepG 2 cells treated with human cytokines

Hepatocyte Treatment	Protein Synthesis (% Control)				
	HPT	FBG	CER	α_1PI	α_1Achy
A	407	279	120	127	317
B	228	194	118	113	288
C	100	114	113	108	121
D	169	138	146	130	300
E	230	193	115	120	264

(A), α_1Achy: cat G complexes; (B), α_1Achy; (C), cat G; (D), PMA, (E), α_1Achy*. Results are expressed as % change in comparison with culture media from unstimulated fibroblasts.

by incubating with HNE, indicating a potential function for this enzyme when in a free state. Clearly, this could occur under conditions where all of its controlling inhibitor, α_1PI, was either complexed, proteolytically inactivated or, more likely oxidatively inactivated. Such conditions are almost certainly likely to prevail at the centre of inflammatory lesions. However, normal synthesis of plasma proteins could be regulated by the simple release of proteinases from fibroblasts or other cells which could convert α_1Achy to α_1Achy* and utilize the latter as a constant stimulant for IL-6 production.

We have also recently found that HNE-inactivated α_1Achy can cause the synthesis of IL-8 in fibroblasts. While this data is relatively new it does suggest that the proteolytic inactivation of Serpins may be a mechanism for stimulating neutrophil chemotaxis by this cytokine. Clearly, further investigations are warranted to determine all of the functions of the modified forms of Serpins during inflammatory episodes.

Neutrophil Proteinases and Cell Movement

One of the least understood mechanisms in neutrophil function is that of cell movement. The commonly accepted theory is that this occurs through diapedesis. However, it is unlikely that this type of cell movement can occur when the neutrophil passes through basement membrane. Some data has already been published which suggests that neutrophils have receptors on their cell surfaces for both HNE and cat G.[24] Indeed, it has been postulated that the presence of such enzymes would allow the cell to degrade connective tissue proteins as it moved towards its target during a chemotactic response. In this laboratory we have approached the problem in two ways. First, we have determined that the carbohydrate structures of the isozyme forms of HNE and cat G are highly varied so that at least some of the forms can be secreted (i.e., they contain secretory carbohydrate side-chains).[6,9] We believe that under certain types of stimulation such secretory forms of HNE and cat G are released from the cell and re-bind on its surface. However, it is doubtful that this occurs through specific receptors. Rather it is more likely that this is simply a series of ionic interactions between the acidic sialic acid residues on cell-surface glycoproteins and the very basic arginine residues of cat G and HNE. Indeed, preliminary data from our laboratory (N. Bangalore, unpublished observations) are strongly in favour of such a hypothesis since basic proteins such as histones can block membrane binding of HNE and cat G after cell stimulation. Significantly, there is only weak binding of myeloperoxidase, and this can easily be explained by its lower isoelectric point.

Second, we have recently noted that α_1Achy*, as well as α_1Achy/cat G complexes, can stimulate the production of IL-8 by fibroblasts. Since it is well known that IL-8 is a chemotactic cytokine[25], we have reasoned that a second role of neutrophil proteinases involves the synthesis of IL-8 which acts to stimulate neutrophil movement towards its target. In this respect we should also point out that data is already available which shows that proteolytically modified Serpins are, themselves, direct chemotactic factors for neutrophils.[26,27] The chemotaxis induced by these proteins is not accompanied by neutrophil degranulation and oxidant burst, and this may be very important in the late stages of inflammation when neutrophils are involved in the healing process.

Final Remarks

From this short description of the likely roles of neutrophil proteinases in inflammation, it is clear that total inhibition of their activities could result in the shutdown of a series of important biochemical reactions. Yet, substantial efforts are being made to develop synthetic HNE inhibitors to reduce tissue damage when regulating Serpin levels are dangerously low.

Clearly, this is a serious dilemma. Yet it can be handled if the synthetic organic chemist is willing to develop oxidation-sensitive inhibitors. Obviously, the methionine residue at the active centre of α-1-PI is present to allow the regulation of HNE inhibitory activity by oxidizing agents. One can easily foresee how a cell must be able to move through a sea of plasma inhibitors towards its target. If HNE is required on the cell surface to degrade proteins during chemotaxis it must not be complexed. This is where oxidation comes into place, myeloperoxidase-derived compounds providing a halo of oxidizing agents around the cell which inactive local α-1-PI. This also allows free HNE to degrade α-1-Achy, not only keeping cat G from being complexed but also providing a source for IL-6 and IL-8 synthesis by fibroblasts and macrophages near the inflammatory site. If synthetic compounds are going to be used to augment HNE-inhibitory activity they will have to be oxidation sensitive to allow neutrophil migration, α-1-Achy* production and, ultimately, the production of acute phase proteins.

The mechanism by which acute phase protein production occurs results in the early rise in α-1-Achy levels followed much later by that of α-1-PI. Under the scenario described above this would seem quite logical. First, α-1-Achy levels rise to guarantee a source of α-1-Achy: cat G complexes and α-1-Achy*. These compounds are then utilized for the production of IL-6 and IL-8 so that both acute phase protein production and chemotaxis can occur. Later, as the levels of α-1-PI rise there is a shutdown in α-1-Achy production since HNE regulation by α-1-PI is now restored. This now turns off both IL-6 and IL-8 production and both neutrophil migration and acute phase production are turned off. Whether this is the correct scenario remains to be proven. However, it seems clear that the role of neutrophil proteinases does not just involve the degradation of proteins ingested by neutrophils. Rather, they are more likely to play sophisticated roles in the more important aspects of the inflammatory process.

References

1. Travis J., Salvesen G.: Human Plasma Proteinase Inhibitors. Ann. Rev. Biochem. 1983; 83: 655-709

2. Beatty K., Bieth J., Travis J.: Kinetics of Association of Serine Proteinases with Native and Oxidized Alpha-1-Proteinase Inhibitor and Alpha-1-Antichymotrypsin. J. Biol. Chem. 1980; 255: 3931-3936

3. Jeppson J.: Amino acid Substitution (Glu to Lys) in Alpha-1-Antitrypsin PiZ. FEBS Letters 1976; 65: 195-197

4. Laurell C.B., Eriksson S.: The Electrophoretic Alpha-1-Globulin Pattern in Alpha-1-Antitrypsin Deficiency. Scand. J. Clin. Lab. Invest. 1963; 15: 132-140

5. Janoff A., Carp H.: Possible Mechanisms of Emphysema in Smokers. Am. Rev. Respir. Dis. 1979; 116: 65-72

6. Travis J., Dubin A., Potempa J., Watorek W., Kurdowska A.: Neutrophil Proteinases: Caution signs in designing inhibitors against enzymes with possible multiple functions. Ann. N.Y. Acad. Sci. 1991; 624: 81-86

7. Farley D., Travis J., Salvesen G.: The human neutrophil elastase gene: analysis of the nucleotide sequence. Hoppe-Seyler's Z. Physiol. Chem. 1989; 370: 737-744

8. Loebermann H., Tokuoka R., Deisenhofer J., Huber R.: Human Alpha-1-Proteinase inhibitor: Crystal structure analysis of two crystal modifications, molecular modeling, and preliminary analysis of the implications for function. J. Mol. Biol. 1984; 177: 531-556

9. Sinha S., Watorek W., Karr S.S., Giles P..J., Bode W., Travis J.: Primary structure of human neutrophil elastase. Proc. Natl. Acad. Sci. USA 1987; 84: 2228-2232

10. Johnson D., Travis J.: Structural evidence for methionine at the reactive site of human Alpha-1-proteinase inhibitor. J. Biol. Chem. 1978; 253: 7142-7145

11. Beatty K., Matheson N., Travis J.: Kinetic and chemical evidence for the inability of oxidized

alpha-1-proteinase inhibitor to protect lung elastin from elastolytic degradation. Hoppe-Seyler's Z. Physiol. Chem. 1984; 365: 731-736

12. Matheson N., Janoff A., Travis J.: Enzymatic oxidation of alpha-1-proteinase inhibitor in abnormal tissue turnover. Mol. Cellular Biochem. 1982; 45: 65-77

13. Morii M., Travis J.: Amino acid sequence at the reactive site of human alpha-1-antichymotrypsin. J. Biol. Chem. 1983; 258: 12749-12753

14. Carrell R., Travis J.: Alpha-1-Antitrytpsin and the Serpins: Variation and Countervariation. TIBS 1985; 10: 20-24

15. Kress L., Kurecki T., Chan S.K., Laskowski M.Sr.: Characterization of the inactive fragments resulting from limited proteolysis of human alpha-1-proteinase inhibitor by Crotalus Adamanteus proteinase II. J. Biol. Chem. 1979; 254: 5317-5320

16. Banda M., Sinha S., Travis J.: Inactivation of human alpha-1-Proteinase inhibitor by macrophage elastase. J. Clin. Invest. 1987; 79: 1314-1318

17. Boudier C., Holle C., Bieth J.: Stimulation of the elastolytic activity of leukocyte elastase by leukocyte cathepsin G. J. Biol. Chem. 1981; 256: 10256-10258

18. Reilly C.F., Funkunaga Y., Powers J.C., Travis J.: Effect of neutrophil cathepsin G on elastin degradation by neutrophil elastase. Hoppe-Seyler's Z. Physiol. Chem. 1984; 365:1131-1135

19. Roughley P., Barrett A.J.: The degradation of cartilage proteoglycans by tissue proteinases. Proteoglycan structure and its susceptibility to proteolysis. Biochem. J. 1977; 167: 6229-6637

20. Odeberg H., Olsson I.: Microbicidal mechanisms of human granulocytes: synergistic effects of granulocyte elastase and myeloperoxidase or chymotrypsin-like cationic protein. Infect. Immun. 1976; 14: 1276-1283

21. Bangalore N., Travis J., Onunka V., Pohl J., Shafer W.M.: Identification of the primary antimicrobial domains in human neutrophil cathepsin G. J. Biol. Chem. 1990; 265: 13584-13588

22. Gauldie J., Northemann W., Fey G.H.: IL-6 functions as an exocrine hormone in inflammation. Hepatocytes undegoing acute phase responses require exogenous IL-6. J. Immunol. 1990; 144: 3804-3808

23. Kurdowska A., Travis J.: Acute phase protein stimulation by alpha-1-antichymotrypsin/cathepsin G complexes. Evidence for the involvement of interleukin-6. J. Biol. Chem. 1991; 265: 21023-21029

24. Dwenger A., Tost P., Hole W.: Evaluation of elastase and alpha-1-proteinase inhibitor-elastase uptake by polymorphonuclear leukocytes and evidence of an elastase-specific receptor. J. Clin. Chem. Clin. Biochem. 1986; 24: 299-308

25. Thornton A., Strieter R., Lindley I., Baggiolini M., Kunkel S.: Cytokine-induced gene expression of a neutrophil chemotactic factor/IL-8 in human hepatocytes. J. Immunol. 1990; 144: 2609-2613

26. Banda M., Rice A., Griffin G., Senior R.: Alpha-1-proteinase inhibitor is a neutrophil chemoattractant after proteolytic inactivation by macrophage elastase. J. Biol. Chem. 1988; 263: 4481-4484

27. Banda M., Rice A., Griffin G., Senior R.: The inhibitor complex of human alpha-1-proteinase inhibitor and human leukocyte elastase is a neutrophil chemoattractant. J. Exp. Med. 1988; 167: 1608-1615

7. Kinetics of the Interaction of Human Leucocyte Elastase with Protein Substrates: Implications for Enzyme Inhibition

A. BAICI

Department of Rheumatology, University Hospital, Zurich, Switzerland

Introduction

Human leucocyte elastase (EC 3.4.21.37) is possibly one of the most destructive enzymes in the body, having the ability to degrade many components of the extracellular matrix such as insoluble collagens type I and II,[1] type III collagen,[2] type IV collagen,[3,4] proteoglycans[1,5] and elastin.[6,7] Other natural substrates degraded by leucocyte elastase are the four human immunoglobulin G subclasses,[8] immunoglobulin M[9] and the cell adhesion molecule fibronectin.[10] For these reasons leucocyte elastase has been associated with pathological states characterized by an abnormal degradation of connective tissue, in particular with pulmonary emphysema,[11,12] rheumatoid arthritis,[13] clotting disorders and other inflammatory processes.[14] An attractive approach for the treatment of emphysema[15-17] and other pathological states characterized by the loss of the structural elements of the extracellular matrix is the use of low molecular mass synthetic inhibitors of endopeptidases. However, despite the existence of a large number of endopeptidase inhibitors that are very active *in vitro*, very few of them are able to exert beneficial effects *in vivo* through inhibition of proteolysis. The manifold reasons for this failure, which are essentially due to the particular nature of the enzymes and their target substrates in the extracellular matrix, such as collagen, proteoglycans and elastin, have been discussed.[18]

The degradation of elastins by elastases of various sources has been the subject of detailed study aimed at showing the importance of electrostatic and hydrophobic enzyme/substrate interactions for elastin solubilization.[19-23] Knowing the properties of a given elastase/elastin system is a necessary prerequisite for investigating the action of inhibitors of elastolysis. A frequent handicap is that free and elastin-bound elastases may be differently susceptible to inhibition by both naturally-occurring and synthetic inhibitors, since many of them are unable to interact with the enzyme once it is tightly bound to the elastin substrate.[24-32]

The formation of an enzyme-inhibitor complex is not always a rapid event completed within the time scale of diffusion-controlled processes. Several substances behave as slow-binding or slow, tight-binding enzyme inhibitors.[33-38] In order for an inhibitor to be physiologically significant, it must be able to bind the target enzyme at a rate greater than that of the natural substrates, otherwise significant turnover would occur before equilibration of the enzyme-inhibitor complex.[18]

The *in vivo* significance of kinetic constants of protein endopeptidase inhibitors was first discussed by Bieth.[39,40] According to this theory, a 'delay time of inhibition' can be calculated from the known kinetic constants for a given enzyme/inhibitor system. In order to predict the physiological significance of an inhibitor, the extent of substrate degradation during the delay time of inhibition must be calculated.

The present contribution constitutes a reappraisal of these concepts and describes a practical approach to them. The serine endopeptidase human leucocyte elastase is chosen as an example to illustrate the binding of the enzyme to representative soluble and insoluble macromolecular substrates. Depending on the substrate, this binding process can be slow or fast and its characteristics allow the formulation of hypotheses on the practical use of elastase inhibitors.

Materials and Methods

The necessary experimental details were described in a preceding paper.[41] Further theoretical and practical information related to the topic treated here can be found elsewhere.[42-47]

Theory and Models

Interaction of elastase with macromolecular substrates

The following models consider an elastolytic enzyme (E) acting on two substrates present simultaneously. One of the substrates is a soluble oligopeptide (S) while the other is a macromolecular substrate (M), which may, or may not, be soluble.[41] The catalytic site of E can either be occupied by S or M. It is assumed that the binding of S to E is fast enough with respect to the catalytic step for the oligopeptide substrate, k^S_{cat}, whereas the binding of M to E can either be fast or slow according to the two mechanisms shown in Scheme 1.

If S is a fluorogenic or chromogenic oligopeptide, the products of S and M (P and Q, respectively) can always be distinguished from each other so that the absorbance or the fluorescence of P can be used to monitor the progress of the reaction, while

Fig. 1.

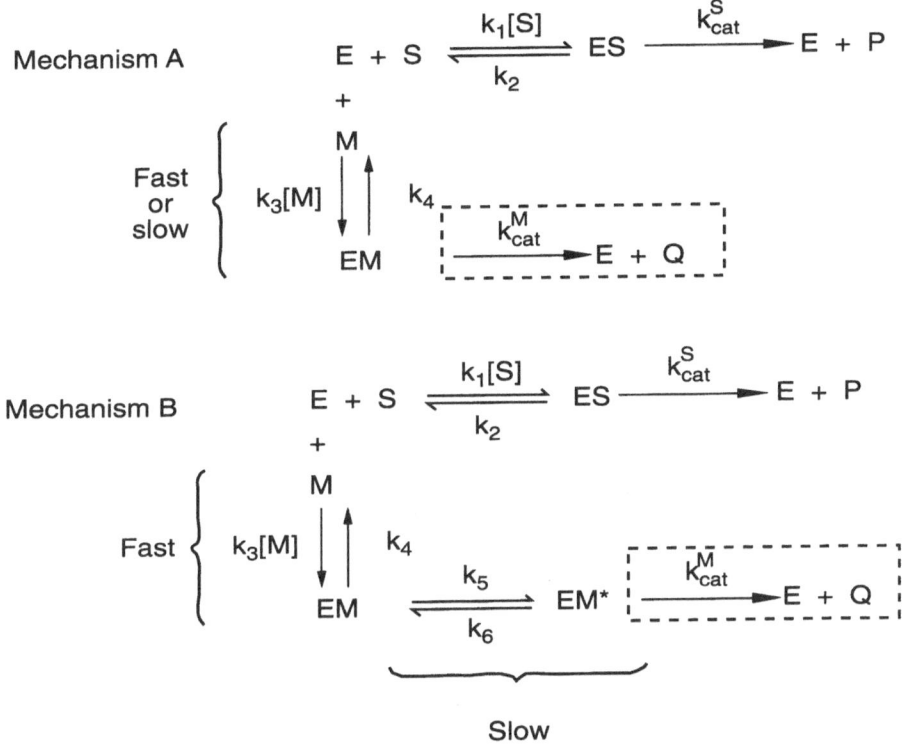

the product Q generated from the macromolecular substrate is deprived of a spectroscopically measurable signal. For this reason S will also be referred to as the 'reporter substrate'. In such a situation, M can be formally treated as a competitive inhibitor of S, and measurement of an 'inhibition constant' for M actually yields the Michaelis constant for this competing substrate.[48] Both of the mechanisms shown in Fig. 1 are reminiscent of the models proposed for slow-binding inhibition.[35-38] If the rate of equilibration between E and M in mechanism A is fast, then the velocity equation for classical, fully competitive inhibition can be used for describing the kinetic properties of the system as follows:[48]

$$v_i = v_o \ \frac{1 + [S]/K^S_m}{1 + [S]/K^S_m + [M]/K^M_m} \qquad (1)$$

Where $[M]/K^M_m$ substitutes the more familiar term $[I]/K_i$ for the reasons explained above. If the proteolytic reaction, k^M_{cat}, in mechanisms A and B is much slower than all other steps, then the boxed portions of mechanisms A and B can be neglected. In this case the term $[M]/K^M_i$ can be substituted for $[M]/K^M_m$ in Eqn. 1 (i.e. M behaves as a classical competitive inhibitor with $K^M_i = k_4/k_3$). The integrated rate equation for progress curves according to both mechanism A (with slow equilibration) and mechanism B in Scheme 1 is then given by:[36]

$$[P] = v_s t + (v_z - v_s)(1 - e^{-\lambda t})/\lambda \qquad (2)$$

with v_z and v_s representing the velocities at zero time and at steady state, respectively. The product concentration [P] approaches exponentially to a linear steady-state with an apparent pseudo-first-order constant, λ. Mechanisms A and B (Scheme 1) are distinguished by different expressions of λ:

$$\text{Mechanism A:} \qquad \lambda = \frac{k_3}{1 + \sigma}[M] + k_4 \qquad (3)$$

$$\text{Mechanism B:} \qquad \lambda = k_6 \left[\frac{1 + \sigma + \dfrac{k_3(k_5+k_6)}{k_4 k_6}[M]}{1 + \sigma + \dfrac{k_3}{k_4}[M]} \right] \qquad (4)$$

Where $\sigma = [S]/K^S_m$ Eqns. 3 and 4 are valid only if the boxed parts in Mechanisms A and B can be ignored, otherwise the expressions would contain also the contributions of the boxed pathways and would look more complicated. As shown below, this occurs in practice: Under the experimental conditions employed, the true proteolytic reaction occurs at a rate much slower than that of the other steps. The limiting values of λ for $[M] \to 0$ and $[M] \to \infty$ are as follows:

Mechanism A: $\lambda \to k_4$ for $[M] \to 0$ $\lambda \to \infty$ for $[M] \to \infty$
Mechanism B: $\lambda \to k_6$ for $[M] \to 0$ $\lambda \to k_5 + k_6$ for $[M] \to \infty$

In Eqn. 2 v_z coincides with the velocity, v_o, in the absence of inhibitor (in this case M) and is independent of its concentration only in the case of mechanism A, whereas it is a hyperbolic function of the inhibitor concentration in the case of mechanism B.[36] The two variants of mechanism A (slow or fast equilibration) and mechanisms A and B (slow equilibration) can be distinguished by applying the following criteria:

1. Form of the reaction profile, i.e. linear from the very beginning or curved;
2. Dependence of v_z upon [M], i.e. linear or hyperbolic;
3. Dependence of λ upon [M], i.e. linear or hyperbolic.

The symbol [M] used above represents the concentration of the macromolecular substrate in moles 1^{-1} and is valid for soluble substrates. For an insoluble substrate [M] represents the number of enzyme binding sites, and the rate constants for complex formation and dissociation shown in Scheme 1 represent rates of adsorption and desorption of the enzyme on/from the substrate. The kinetic models just examined will be used only with the purpose of calculating the values of the apparent constant λ (with dimension $= s^{-1}$) for either [M] $\rightarrow \infty$ (i.e. the enzyme is saturated by M) or [M] $\rightarrow$ 0, thus making the calculation independent on the physical meaning of [M].

Integrated Henri-Michaelis-Menten Equation
The determination of K_m for a macromolecular substrate in the case of Mechanism A with fast equilibration of the EM complex (Scheme 1 and Eqn. 1) was discussed in the preceding section while the ratio k_{cat}/K_m can be obtained using the integrated form of the Henri-Michaelis-Menten equation (Eqn. 5). The reader is referred to the specific literature for full details.[49,50]

$$\frac{1}{t} \ln \frac{[S]_o}{[S]} = -\frac{1}{K_m} \frac{[S]_o - [S]}{t} + \frac{V}{K_m} \quad (5)$$

Eqn. 5 is just one of several possible forms of the integrated rate equation and is generally valid, provided the reaction is not inhibited by either the substrate or by extraneous inhibitors.[49] Under conditions where $[S]_o \ll K_m$ the integrated rate equation assumes the following simple form:

$$\ln \frac{[S]_o}{[S]} = \frac{V}{K_m} t \quad (6)$$

In practice, experimental data can first be plotted according to Eqn. 6. A straight line passing through the origin will indicate that the condition $[S]_o \ll K_m$ applies, otherwise a plot according to Eqn. 5 can be used.

The ratio V/K_m can be obtained from both equations. If the total enzyme concentration, $[E]_o$, is known, dividing V/K_m by $[E]_o$ yields k_{cat}/K_m.

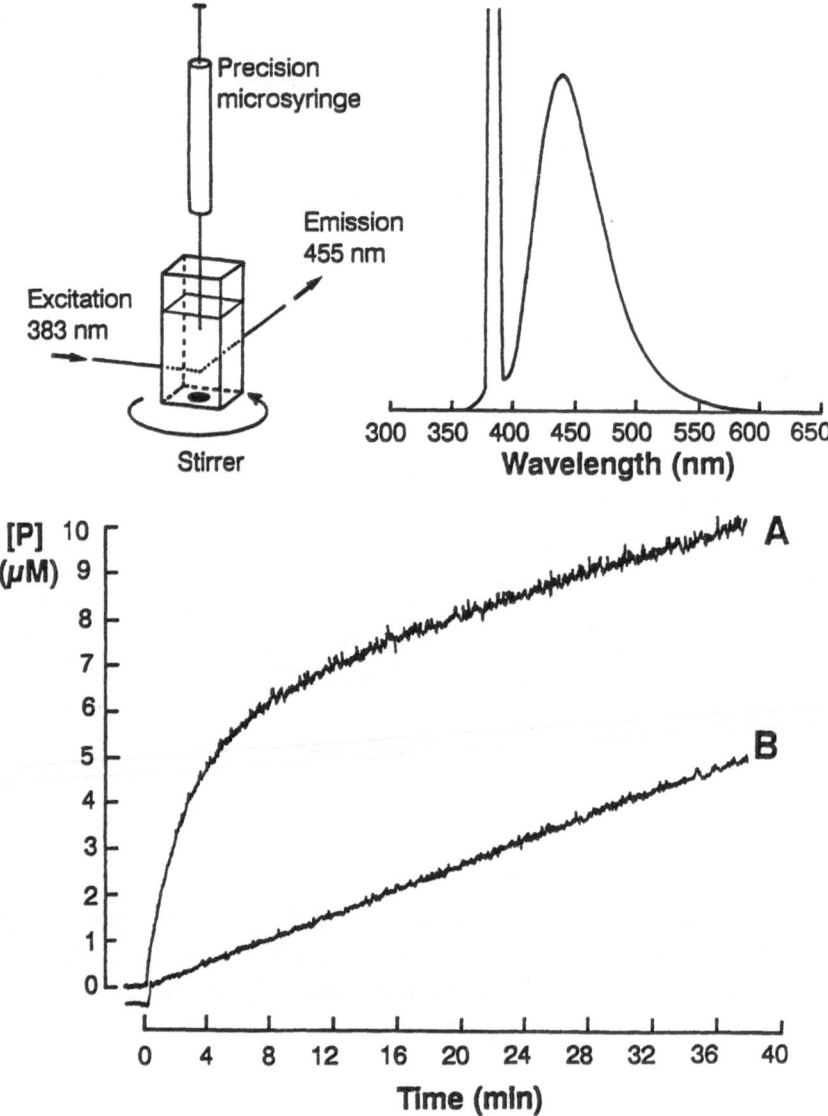

Fig. 1. Binding of human leucocyte elastase to insoluble elastin from bovine neck ligament. The reacting suspension in a fluorescence cuvette contained the reporter substrate Suc-Ala-Ala-Pro-Val-NMec (0.81 mM), elastin (2.36 mg/ml) and elastase (19 nM). The upper left figure shows how additions of either enzyme or substrate solution were made to the continuously stirred suspension without necessarily opening the sample compartment of the fluorimeter. The upper right figure shows the fluorescence emission spectrum of the reaction product, 7-amino-4-methylcoumarin at a concentration of 5 μM, in the presence of 3.5 mg/ml of elastin. The light-scattering peak at the excitation wavelength (383 nm) was kept within reasonable limits using a narrow excitation bandwidth (0.5 or 1nm). The progress curves were obtained by adding enzyme to preincubated reporter substrate and elastin (A) or by adding substrate to preincubated enzyme and elastin (B). 50 mM Na^+/K^+ phosphate buffer, pH 7.4, containing 0.1% (w/v) of Triton X-100, 37°C.

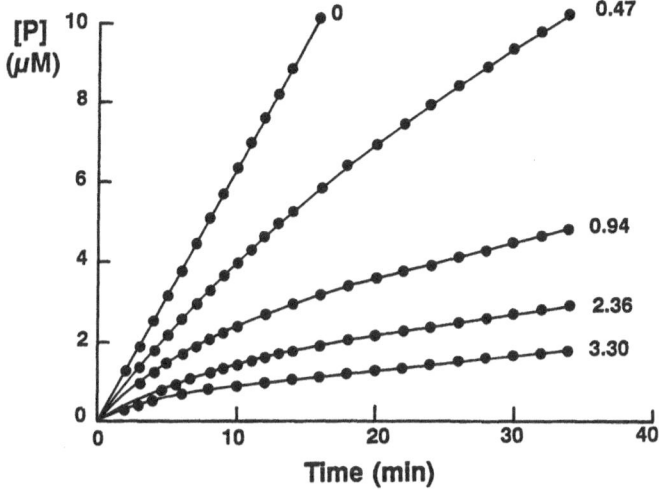

Fig. 2. Progress curves obtained as in Fig. 1A. The reporter substrate was Acetyl-Ala-Ala-Pro-Ala-NMec (0.24 mM) and the reactions were started by adding elastase to a final concentration of 19 nM. The amounts of bovine neck ligament elastin (mg/ml) are indicated by numbers. Points are experimental and lines are best-fit curves to Eqn. 2. For nonlinear regression analysis 30-40 points were sampled from curves as that in Fig. 1A, but only a few of them are shown here for clarity.

Results

Interaction of leucocyte elastase with elastins

The adsorption of human leucocyte elastase to elastin from bovine neck ligament and porcine aorta was investigated using the models described by Scheme 1 and Eqns. 2-4.

Either Suc-Ala-Ala-Pro-Val-NMec or Acetyl-Ala-Ala-Pro-Ala-NMec were used as soluble, fluorogenic reporter substrates. The elastins were added as a fine powder to give suspensions with 0.47-3.30 mg of elastin per ml.

Figure 1 shows a representative experiment and explains further details of the technique. The reaction profiles obtained by adding elastase to a mixture containing all other reactants consisted of an exponential rise of product concentration, lasting several minutes, followed by a linear steady-state (Fig. 1A).

On the other hand, if elastase was previously incubated with elastin and the reaction started by adding the soluble substrate, the reaction profiles were linear from the beginning with slopes identical to those obtained at steady-state with the other pre-incubation (Fig. 1B).

Reaction profiles as that shown in figure 1A were used for calculating the apparent pseudo-first-order constant λ, as well as v_s and v_z (Eqn. 2): 30-40 points were sampled from each experimental curve and the data were fitted by nonlinear regression to Eqn. 2, as illustrated in figure 2 for elastin from bovine neck ligament.

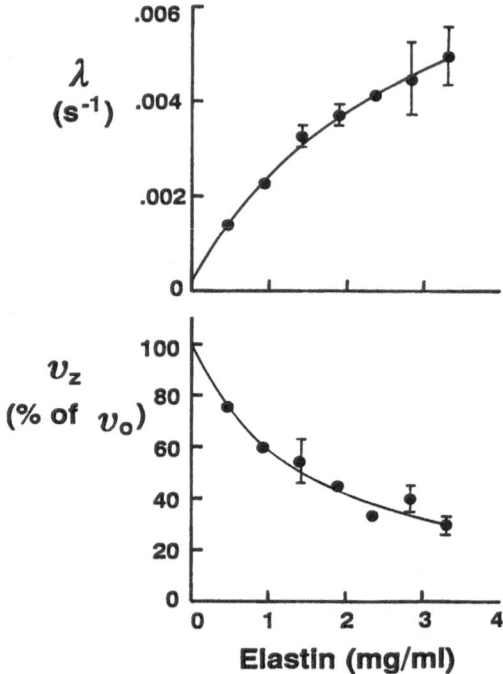

Fig. 3. Variation of λ and v_z as a function of the elastin amount. Data, obtained from Fig.2, were weighted according to the squared inverse of their standard errors and fitted to Eqns. 4 and 1 for λ and v_z, respectively, with v_z substituting v_i in the latter case. Bars indicate $\pm$ S.D.: for some points this is within the size of the symbols.

In separate experiments it was demonstrated that elastolysis was very slow during the exponential phase of the reaction, thus justifying the use of Eqns. 3 and 4 for analysing the dependence of λ upon the amount of elastin present.[41]

The dependence of both λ and v_z upon the amount of elastin from bovine neck ligament and from porcine aorta was hyperbolic (see Fig. 3 for bovine elastin). The individual values of λ depended upon the size of the elastin particles used. However, the extrapolated values at infinite and zero elastin amount were independent on particle size. Since only these extrapolated values are actually needed, the use of an elastin powder of definite granularity was superfluous.

The data in figures 1-3 suggest that the adsorption of human leucocyte elastase to elastin is a time-dependent phenomenon conforming to Mechanism B in Scheme 1. Elastin from porcine aorta behaved qualitatively as elastin from bovine neck ligament, albeit with different kinetic parameters, as summarized in Table I.

Interaction of leucocyte elastase with soluble plasma proteins

While in the preceding section an insoluble protein substrate of elastase was considered, the behaviour of two soluble proteins, bovine serum albumin and a human monoclonal immunoglobulin, will now be analysed. Albumin was chosen as a generic protein, while IgG_3 may represent a true substrate in pathological conditions for leucocyte elastase, whose properties have been previously characterized.[8]

Table I. Limiting kinetic constants for the binding of human leucocyte elastase to insoluble elastin

Constant	Elastin from bovine neck ligament	Elastin from porcine aorta
k_6 (s^{-1})	$1.9 \pm 1.2 \times 10^{-4}$	$5.2 \pm 4.4 \times 10^{-4}$
$k_5 + k_6$ (s^{-1})	$9.5 \pm 1.2 \times 10^{-3}$	$2.1 \pm 0.2 \times 10^{-3}$
t $^{1}/_{2}$ for binding (min)	1.2	5.5
$\simeq$ 99% binding (min)[a]	8.4	38.5

[a]Time required for almost total adsorption of elastase on elastin (7 x t $^{1}/_{2}$) under saturating conditions. k_6 and ($k_5 + k_6$) represent the extrapolated values of λ according to Mechanism B (Scheme 1 and Eqn. 4) when the number of elastase binding sites tend to zero or infinity, respectively. Best-fit values ±S.E. from non linear regression analysis.

Reaction progress curves for the hydrolysis of synthetic peptide substrates by leucocyte elastase in the presence of bovine serum albumin and human IgG3 were recorded both fluorimetrically and spectrophotometrically using the same technique described in the preceding section. Independently of whether elastase was previously incubated with the macromolecular substrates or not, reaction profiles were always linear from the very beginning in the whole concentration range of reactants explored (data not shown). Therefore, kinetic analysis was performed using the slopes of the straight lines obtained for various concentrations of the reporter substrate and of the protein substrates. Preliminary estimates of the kinetic parameters and the type of 'inhibition' caused by the macromolecular substrates were guessed from the specific velocity plot, a graphical method that enables the unequivocal diagnosis of both linear and hyperbolic inhibitors[42], as illustrated in Fig. 4 for bovine serum albumin.

The set of straight lines obtained for three reporter substrate concentrations and four albumin concentrations converged to $v_o /v_i = 1$ on the right ordinate axis. The secondary plot, shown as an inset in figure 4, had an ordinate intercept of 1. These features are characteristic of linear, i.e. fully competitive inhibition, for which the specific velocity equation and the equation for the secondary plot assume the following particular forms, respectively:[42]

$$\frac{v_o}{v_i} = -\frac{[I]}{K_i} \frac{\sigma}{1+\sigma} + 1 + \frac{[I]}{K_i} \qquad (7)$$

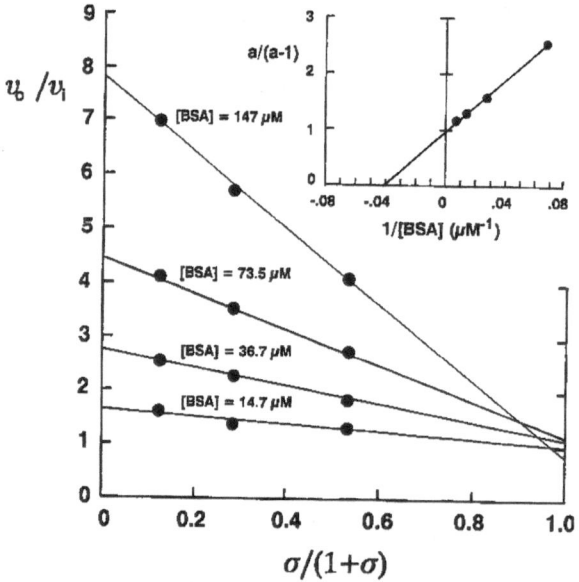

Fig. 4. Specific velocity plot for the 'inhibition' of human leucocyte elastase by bovine serum albumin (BSA). Reporter substrate = Suc-Ala-Ala-Val-p-nitroanilide. The inset shows a replot according to eqn. 8. Buffer and temperature as in Fig. 1.

$$\frac{a}{a - 1} = K_i \frac{1}{[I]} + 1 \qquad (8)$$

In these equations $\sigma = [S]/K_m$ and 'a' represent the left-ordinate intercept of the straight lines in the specific velocity plot. From the inset plot in Fig. 4, a K_i value

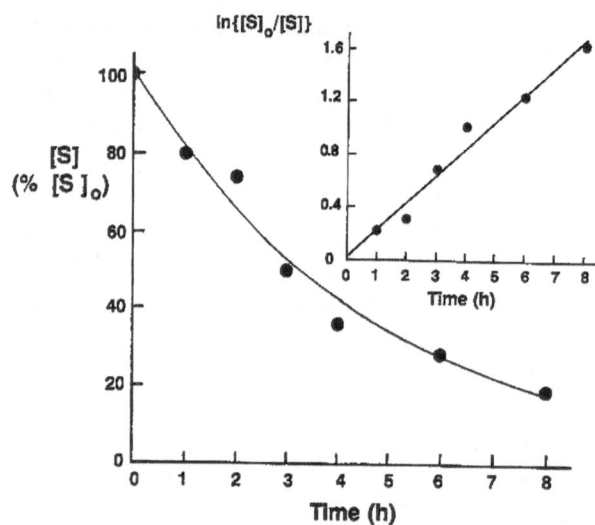

Fig. 5. Cleavage of human monoclonal IgG3 by leucocyte elastase.[8] Data are expressed as the percentage of the IgG3 substrate (S) remaining after specified incubation times. The slope of the inset plot according to eqn. 6 gives V/K_m.

(having actually the meaning of K_m) of 2.5×10^{-5} M could be guessed for bovine serum albumin. After this preliminary diagnosis, refinement of the kinetic constants was obtained by nonlinear regression fitting of data to Eqn. 1 with the following results: $K_m = 2.1 \pm 0.2 \times 10^{-5}$ M and $3.8 \pm 0.3 \times 10^{-5}$ M for bovine serum albumin and IgG3, respectively.

Measurement of k_{cat}/K_m for protein substrates

This procedure, based on results described in detail elsewhere[8], is shown here for IgG3. The protein was incubated with elastase at 37°C and pH 7.4; aliquots were withdrawn at time intervals and the reaction stopped by adding sodium dodecyl sulfate to a final concentration of 2% (w/v) and boiling for 2 min. The digestion products were analysed by slab-gel polyacrylamide gel electrophoresis in the presence of sodium dodecyl sulphate. After staining of the protein bands, the gels were scanned and the amounts of undegraded IgG3 and of the reaction products (Fch and Fab) were deduced from the area of the integrated peaks. The proportion of undergraded IgG3 remaining after given incubation times is shown in figure 5 together with a plot according to Eqn. 6. The straight line passing near the origin of the axes suggests that first-order conditions ($[S]_o \ll K_m$), may apply. Even if true first-order conditions should not apply, the slope of the plot in the inset of Fig. 5 yields a reliable estimate of the V/K_m ratio, as can be judged by inspection of Eqns. 5 and 6. Dividing V/K_m by the titrated enzyme concentration yields $k_{cat}/K_m = 63$ M^{-1} s^{-1}. From the value of $K_m = 3.8 \cdot 10^{-5}$ M, determined independently in the preceding section, it follows that $k_{cat} = 2.4 \times 10^{-3}$ s^{-1}.

Inhibition of elastase in the presence of protein substrates

The progress-curve kinetic method described above (Fig. 1) allowed the direct examination of the behaviour of elastase inhibitors in the presence of both soluble and insoluble macromolecular substrates of the enzyme. In this case the reacting mixture contained the reporter fluorogenic substrate, the protein substrate, an inhibitor and the enzyme. The order of additions of the reactants and various pre-incubations could be combined in various ways to gather information on the efficiency of the inhibitors. This concept is illustrated by an example in figure 6 for the leucocyte elastase inhibitor eglin c. The reacting mixture contained initially the reporter substrate and bovine neck ligament elastin and the reaction was initiated by adding a small volume of an elastase solution. While the reaction was in the exponential phase, a small volume of an eglin solution was rapidly added. As can be seen in figure 6, eglin c could efficiently inhibit the enzyme and the reaction trace after addition of eglin was practically superimposable on that in the presence of eglin alone. This suggests that eglin c is able to efficiently compete with elastin for the binding of elastase. An identical behaviour was found by adding eglin to the reacting mixture at any time during the progress of the reaction. The same result was obtained with bovine serum albumin as the macromolecular substrate. In this case

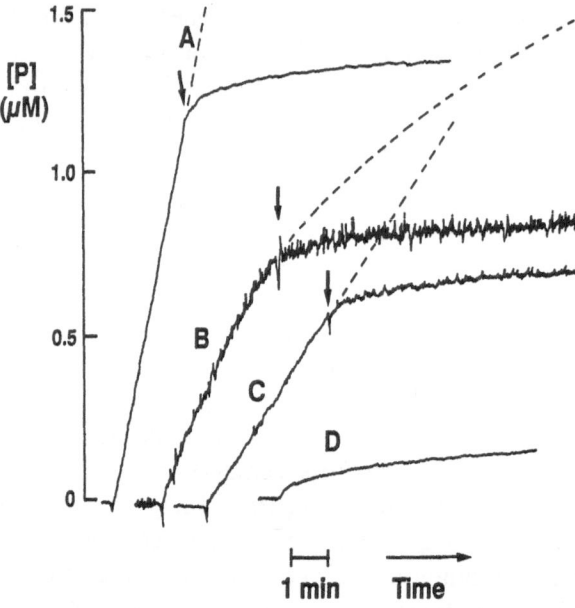

Fig. 6. Inhibition of human leucocyte elastase by eglin c in the presence of protein substrates. The progress curves were obtained by adding elastase (19 nM) to solutions (or suspensions) containing Acetyl-Ala-Ala-Pro-Ala-NMec (0.24 mM) as the reporter substrate. Control with reporter substrate alone (A); reporter substrate plus 1.4 mg/ml of bovine neck ligament elastin (B); reporter substrate plus 28 μM bovine serum albumin (C). A small volume of eglin c (final concentration = 21 nM) was rapidly added at the times indicated by an arrow. Curve D shows the inhibition caused by 21 nM eglin in the presence of reporter substrate alone. In the absence of eglin c curves A, B and C proceeded as shown by the dashed traces. Buffer and temperature as in Fig. 1.

the reaction profile was linear, i.e. at steady-state, from the very beginning of the reaction, and addition of eglin efficiently converted the enzyme to an inactive complex.

Discussion

Adsorption of elastase on elastin and elastolytic reaction

Robert et al.[20] described some years ago the kinetics of elastolysis of iodine-labelled elastin by pancreatic elastase and proposed a three-step mechanism.

1. The enzyme is first adsorbed on a limited surface of the insoluble substrate.
2. The adsorption phase is followed by a structural rearrangement of elastin, probably due to limited proteolysis by the preadsorbed enzyme, which induces an unfolding of peptide chains stabilized by hydrophobic interactions. Unfolding yields a larger surface and a higher local concentration of peptide bonds available to the preadsorbed enzyme.
3. The elastolytic reaction then proceeds at a steady-state rate and desorption of elastase from the substrate is probably controlled by a structural change of the elastic fibre as a consequence of its partial hydrolysis.

The kinetic results shown in the present study agree with the general mechanism of elastolysis of Robert et al. In fact, the adsorption of leucocyte elastase to both bovine neck ligament elastin and porcine aortic elastin occurred according to Mechanism B in Scheme 1 and consisted of three measurable phases: (1) a rapid, reversible recognition of the substrate with complex formation followed by (2) a slow, reversible 'isomerization' of the encounter complex to a modified complex after which (3) the elastolytic reaction proceeded at a steady-state rate. These three steps very probably represent a kinetic verification of the theory of Robert et al. (steps 1-3 discussed above). The slow, reversible step (2) may consist of an unfolding of the elastin structure following the partial hydrolysis of some peptide bonds, which is not necessarily accompanied by release of peptides. This may happen if crosslinking peptide chains are cleaved while leaving the cleaved moieties still bound at their ends to the elastin fibre.

That elastase can dissociate from elastin and bind to another elastin binding site has been unequivocally demonstrated by preadsorbing elastase on ^{125}I-elastin, adding ^{131}I-elastin and showing that the ratio of [131]/^{125}I-peptides in the supernatant increased linearly with time.[20] 'Dissociation' of elastase can either be interpreted as a physical removal of one enzyme molecule from a degraded elastin site followed by a jump of this molecule to another site or as a sliding of the enzyme from a reacted site to an intact site without physical detachment of elastase from its substrate. This concept is illustrated in figure 7. In figure 7A the enzyme is seen to 'jump' from one site to another, while in figure 7B the enzyme moves (crawls) on the elastin substrate in a caterpillar-like fashion. Combined evidence from elastolysis measurements in the presence and absence of various inhibitors favours the caterpillar-like mechanism or a hybrid between the two mechanisms allowing for short-range 'jumps' of the enzyme.

Figure 7C explains why a small-sized substrate (this study) or a small-sized inhibitor[32] can form complexes with elastase without dissociating the enzyme from elastin. In line with recent results by Morrison et al.[32], figure 7D explains how a medium-sized inhibitor, such as eglin c, can inhibit elastase by dissociating the enzyme from the complex with elastin. Finally, figure 7E shows that an inhibitor having a large molecular mass, such as α_1-proteinase inhibitor, may not be able to form complexes with elastase preadsorbed on elastin.[24,26,28,30] However, at a high inhibitor/enzyme molar ratio α_1-proteinase inhibitor has also been shown to dissociate leucocyte elastase from the enzyme-elastin complex.[32] This can probably occur by competition between inhibitor and elastin for the elastase active centre, in the presence of a large excess of inhibitor molecules over enzyme molecules, during occasional short-range 'jumps' of the enzyme on the elastin surface.

The half-time for binding of elastase to elastin under saturating conditions, i.e. with a very large elastin surface available for enzyme binding, depended on the nature of elastin. After about 7 half-times the binding was almost complete and elastolysis occurred at a steady-state rate. Under saturating conditions, leucocyte elastase took about 8 min for total adsorption to elastin from bovine neck ligament

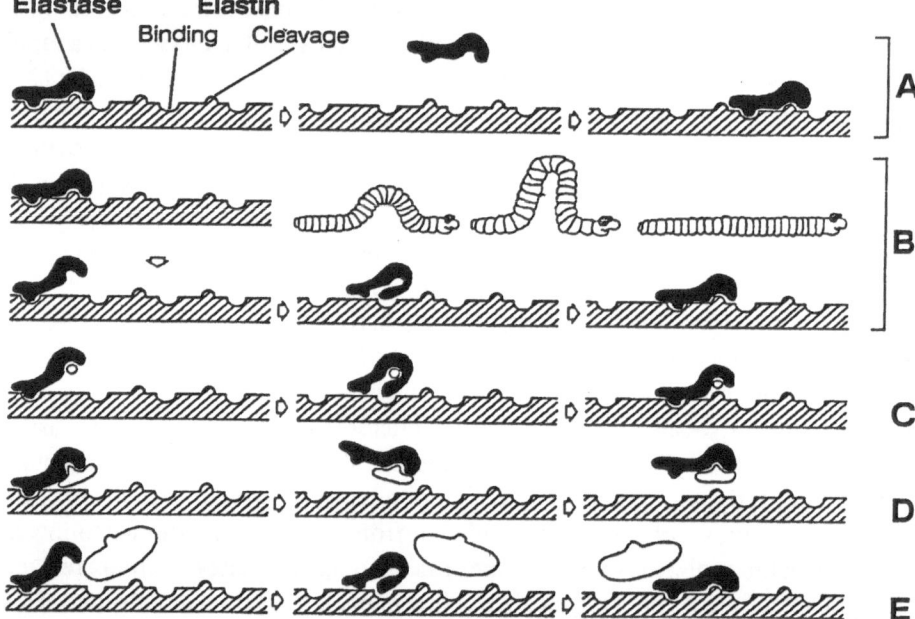

Fig. 7. Hypothetical mechanisms of elastolysis and inhibition of elastases. In A the enzyme hydrolyses an elastin peptide bond and then jumps to another (remote or proximal) site. In B, after hydrolysis, the enzyme crawls on the elastin surface similar to a caterpillar without leaving the elastin surface. The small circle in C represents a low molecular mass inhibitor or substrate. The size of this molecule is not sufficient for dissociating elastase from elastin, but can still occupy the active centre of the enzyme. A medium-sized inhibitor, such as eglin c, can reach the active centre of the enzyme while this is still elastin-bound (D). However, reassociation with elastin is impaired when the enzyme attempts to crawl on the elastin surface. An inhibitor with a larger molecular mass, such as α_1-proteinase inhibitor, cannot reach the active centre of the enzyme crawling to the next binding site (E).

and some 38 min for complete binding to porcine aortic elastin. Clearly, under non-saturating conditions, these times are even longer. In a reaction mixture containing the reporter substrate, elastin and elastase, the slope of the steady-state phase of the reaction was independent of whether elastase had been preadsorbed on elastin or not, while it depended on the amount of elastin available to the enzyme and on the concentration of the reporter substrate, thus suggesting a striking analogy with the kinetics of enzyme-catalysed reactions in the presence of slow-binding inhibitors.[36] However, for these inhibitors the reaction trace obtained by adding the reporter substrate to the preincubated mixture of all other reactants starts with a concave upward lag phase.[36,43] This curvature was not seen when adding S to preincubated E and M (see Scheme 1), thus suggesting that, while the ES complex can be dissociated by M, the EM complex cannot be dissociated by S. The reaction measured as turnover of S into P is likely to occur on either free enzyme, i.e. on enzyme molecules 'jumping' from one elastin binding site to another, or on enzyme molecules bound to elastin, but leaving the active centre accessible to small molecules. A schematic picture of this concept is given in figure 7C.

The caterpillar-like mechanism shown in figure 7 (see in particular Fig. 7B) would predict a sort of noncompetitive relationship between S and M because of the presence of a ternary ESM complex. This apparently contrasts with the proposed mechanism B (Scheme 1), which has a genuine competitive character. In enzyme kinetics, noncompetitive inhibition can either arise if an inhibitor shares the active centre of the enzyme with S or binds to a different site. For the kineticist the two modes of binding are indistinguishable from one another. The assumption made in analysing mechanism B was that the catalytic site of E can be occupied by either M or S, which is still the case in the caterpillar mechanism shown in figure 7. For a slow-binding inhibitor with noncompetitive character, Eqn. 2 is still valid and λ depends linearly on [I].[33,46]

Since the binding of elastase to elastin λ was found to depend hyperbolically upon the amount of elastin present, a mechanism with noncompetitive character can be ruled out.

Soluble proteins as substrates of elastase

While the overall binding of elastase to elastin was a slow process, the enzyme formed very rapidly reversible complexes with bovine serum albumin and human IgG3, considered here as models for soluble plasma proteins. From a formal point of view, the kinetic behaviour of the proteins was identical to that of classical competitive inhibitors (mechanism A with fast equilibration in Scheme 1). Indeed, both albumin and IgG3 behaved as competitive substrates and the inhibition constants determined had the meaning of Michaelis constants (Eqn. 1). However, the definition of a Michaelis constant for a proteolytic reaction is a difficult task. In fact, the hydrolysis of a peptide bond inside a protein generates two new peptides that may represent new substrates. The mechanism is thus complicated by the fact that new potential substrates, each of them probably having a different affinity for the enzyme, are continuously generated during the progress of the reaction. Therefore, one should consider as many individual Michaelis constants as susceptible peptide bonds are present. In addition, the number of substrates increases progressively with time. It is not difficult to imagine that analysing such a system from a kinetic point of view requires hopelessly complex mathematics. However, in the presence of a polypeptide substrate, an endopeptidase usually first attacks a particularly susceptible bond, whereas attack to sites of lower affinity occurs after all of the high-affinity sites have been hydrolysed. This concept has been verified experimentally for the cleavage of IgG3 by leucocyte elastase into Fab and Fch fragments. While Fab represents a final, stable product, Fch is further degraded at a slow rate. However, no degradation of Fch is observed before all of the available IgG3 is first fragmented.[8] Thus, the initial velocity approach used in this study is likely to yield a good estimate of K_m for the hydrolysis of the most susceptible bond on the considered protein substrate. As shown above, the k_{cat}/K_m ratio can be obtained from the integrated rate equation applied to simple experiments, so that the individual kinetic parameters for a protein substrate of an endopeptidase can be calculated.

Inhibition of elastase in the presence of protein substrates

The behaviour of inhibitors of endopeptidases, as investigated *in vitro* using synthetic oligopeptide substrates of the enzymes, may in general not reflect the behaviour of the inhibitors when the enzyme acts on a macromolecular protein substrate. While studies with synthetic substrates are indispensable tools for unravelling the kinetic mechanism and determining the kinetic parameters of inhibition, only the use of the natural substrates allows a reliable evaluation of the possible physiological or pharmacological efficiency of an inhibitor. This is particularly valid for leucocyte elastase in pathological situations, such as pulmonary emphysema, characterized by an excessive degradation of extracellular matrix components. Numerous studies have recognized the fact that inhibitors of elastase are less efficient when using elastin as the substrate instead of oligopeptides, and that some inhibitors are less efficient, or even inefficient, on elastase preadsorbed on elastin.[24-32]

In order for an inhibitor to have physiological or pharmacological significance, it must either bind an enzyme escaping from a pathological cell at a rate greater than the rate of binding of the enzyme to its natural substrates, or it must bind rapidly enough an already present enzyme in order to protect the substrate from excessive turnover. Several inhibitors, either reversible or irreversible, are characterized by slow-binding rates to the enzymes. This concept has been first analysed by Bieth, who also provided a basic theory for predicting the effectiveness of an inhibitor by defining a 'delay time of inhibition'. [39,40]

Bieth's original definition of the delay time of inhibition (DT) has been extended to take into account both reversible and irreversible inhibitors as well as the effect of substrate and the inhibition mechanism:[18]

$$DT \simeq 5/\lambda$$

In this simple expression, λ represents a first-order or a pseudo-first-order constant which describes the exponential phase of binding of an inhibitor to its target enzyme. DT can be calculated knowing the mechanism, the kinetic constants and the inhibitor concentration. A calculation of the amount of substrate degraded during the delay time of inhibition gives an estimate of the inhibitor efficiency. While this concept sounds apparently easy, its practical use is complicated by several problems. In fact, not only the k_{cat}/K_m ratio for the protein substrate must be known, but also the mode of binding of the enzyme to the substrate and the concentration of the enzyme must be known. Furthermore, the k_{cat}/K_m ratio for an insoluble substrate is deprived of physical meaning. In his valid approach, Bieth calculated for the delay time of inhibition a maximum allowable value of 1 to 2 s if neither the concentration of the enzyme nor the kinetic parameters of the enzyme/substrate system are known.[39,40] This limit of 2 s may constitute a handicap for the use of several inhibitors having either unfavourable kinetic constants or insufficient concentrations *in vivo*. However, as shown in this study, considering the individual cases may help in resolving such problems. As far as leucocyte elastase with elastin as the substrate is concerned, it was shown that the process of binding prior to efficient elastolysis requires at least some 8 and 38 min for two different elastin types. Since during the binding phase elastin degradation is very slow, it can be theoretically assumed that the enzyme can be efficiently inhibited even by inhibitors characterized by relatively long delay times for inhibition.

Contrary to the elastase/elastin system, soluble protein substrates of elastase formed very rapidly reversible complexes with the enzyme, from which elastase could readily be dissociated in the presence of inhibitors. In this case, substrate damage will depend on the amount of protein degraded in the absence of inhibitor. After sufficient inhibitor has been made available, one can think of efficient protection against further proteolysis. An approximate calculation of the amount of protein degraded during the delay time of inhibition can be made according to Eqn. 18 in Bieth's paper.[40] As an example we want to calculate the amount of substrate degraded during a hypothetical delay time of inhibition of 1 min supposing an enzyme concentration of 1 μM, which probably represents the highest limit. A substrate with k_{cat}/K_m = 63 $M^{-1}s^{-1}$ will be degraded by 0.076% (this is the case of IgG3/leucocyte elastase); a substrate with k_{cat}/K_m = 517 $M^{-1}s^{-1}$ will be degraded by 0.62% (this is the case of IgG3/cathepsin G[9]); and a hypothetical substrate with k_{cat}/K_m = 5000 $M^{-1}s^{-1}$ will be degraded by 6%. Under the same conditions elastin degradation would be insignificantly small. Thus it is evident that, where possible, consideration of individual systems may better help the prediction of the efficiency of enzyme inhibitors.

Acknowledgments

I wish to thank Mrs. D. Hörler for her skilled assistance with the experimental work and my son Federico for help in drawing Fig. 7.

References

1. Starkey P.M., Barrett A.J., Burleigh M.C.: The degradation of articular collagen by neutrophil proteinases. Biochim. Biophys. Acta 1977; 483: 386-397
2. Mainardi C.L., Hasty D.L., Seyer J.M., Kang A.H.: Specific cleavage of human type III collagen by human polymorphonuclear leukocyte elastase. J. Biol. Chem. 1980; 255: 12006-12010

3. Davies M., Barrett A.J., Travis J., Sanders E., Coles G.A.: The degradation of human glomerular basement membrane with purified lysosomal proteinases: evidence for the pathogenic role of the polymorphonuclear leukocyte in glomerulonephritis. Clin. Sci. Mol. Med. 1978; 54: 233-240

4. Mainardi C.L., Dixit S.N., Kang A.H.: Degradation of type IV (basement membrane) collagen by a proteinase isolated from human polymorphonuclear leukocyte granules. J. Biol. Chem. 1980; 255: 5435-5441

5. Keiser H., Greenwald R.A., Feinstein G., Janoff A.: Degradation of cartilage proteoglycan by human leukocyte granule neutral proteases. A model of joint injury. II. Degradation of isolated bovine nasal cartilage proteoglycan. J. Clin. Invest. 1976; 57: 625-632

6. Janoff A., Scherer J.: Mediators of inflammation in leukocyte lysosomes. IX. Elastinolytic activity in granules of human polymorphonuclear leukocytes. J. Exp. Med. 1968; 128: 1137-1155

7. Galdston M., Levytska V., Liener I.E., Twumasi D.Y.: Degradation of tropoelastin and elastin substrates by human neutrophil elastase, free and bound to alpha2-macroglobulin in serum of the M and Z(Pi) phenotypes for alpha1-antitrypsin. Amer. Rev. Respir. Dis. 1979; 119: 435-441

8. Baici A., Knöpfel M., Fehr K., Skvaril F., Böni A.: Kinetics of the different susceptibility of the four human immunoglobulin G subclasses to proteolysis by human lysosomal elastase. Scand. J. Immunol. 1980; 12: 41-50

9. Baici A., Knöpfel M., Fehr K., Böni A.: Cleavage of human IgM with human lysosomal elastase. Immunol. Lett. 1980; 2: 47-51

10. McDonald J.A., Kelley D.G.: Degradation of fibronectin by human leukocyte elastase. Release of biologically active fragments. J. Biol. Chem. 1980; 255: 8848-8858

11. Snider G.L.: Pathogenesis of emphysema and chronic bronchitis. Med. Clin. North Amer. 1981; 65: 647-665

12. Janoff A.: Elastase in tissue injury. Annu. Rev. Med. 1985; 36: 207-216

13. Barrett A.J.: The possible role of neutrophil proteinases in damage to articular cartilage. Agents Actions 1978; 8: 11-18

14. Fritz H., Jochum M., Duswald K.H., Dittmer H., Kortmann H., Neumann S., Lang H.: Granulocyte proteinases as mediators of unspecific proteolysis in inflammation: a review. In: Goldberg D.M., Werner M. (Eds.): *Selected topics in clinical enzymology*, Vol. 2. Berlin, W. de Gruyter 1984; 305-328

15. Trainor D.A.: Synthetic inhibitors of human neutrophil elastase. Trends Pharmacol. Sci. 1987; 8: 303-307

16. Weinbaum G., Damiano V.V.: Protease inhibitor therapy in emphysema: a promising theory with problems. Trends Pharmacol. Sci. 1987; 8: 6-7

17. Groutas W.C.: Inhibitors of leukocyte elastase and leukocyte cathepsin G. Agents for the treatment of emphysema and related ailments. Med. Res. Rev. 1987; 7: 227-241

18. Baici A.: Criteria for the choice of inhibitors of extracellular matrix-degrading endopeptidases. In: Glauert A.M. (Ed.): *The control of tissue damage*. Amsterdam, Elsevier 1988; 243-258

19. Gertler A.: The non-specific electrostatic nature of the adsorption of elastase and other basic proteins on elastin. Eur. J. Biochem. 1971; 20: 541-546

20. Robert B., Hornebeck W., Robert L.: Cinétique hétérogène de l'interaction élastine-élastase. Biochimie 1974; 56: 239-244

21. Jordan R.E., Hewitt N., Lewis W., Kagan H., Franzblau C.: Regulation of elastase-catalyzed hydrolysis of insoluble elastin by synthetic and naturally occurring hydrophobic ligands. Biochemistry 1974; 17: 3497-3503

22. Kagan H.M., Lerch R.M.: Amidated carboxyl groups in elastin. Biochim. Biophys. Acta 1976; 434: 223-232

23. Lonky S.A., Wohl H.: Regulation of elastolysis of insoluble elastin by human leukocyte elastase: stimulation by lysine-rich ligands, anionic detergents, and ionic strength. Biochemistry 1983; 22:

3714-3720

24. Reilly C.F., Travis J.: The degradation of human lung elastin by neutrophil proteinases. Biochim. Biophys. Acta 1980; 621: 147-157

25. Kueppers F., Abrams W.R., Weinbaum G., Rosenbloom J.: Resistance of tropoelastin and elastin peptides to degradation by α2-macroglobulin-protease complexes. Arch. Biochem. Biophys. 1981; 211: 143-150

26. Hornebeck W., Schnebli H.P.: Leukocyte elastase adsorbed to elastin is incompletely inhibited by α1-proteinase inhibitor. Hoppe Seyler's Z. Physiol. Chem. 1982; 363: 455-458

27. Hornebeck W., Brechemier D., Jacob M.P., Frances C., Robert L.: On the multiplicity of cellular elastases and their inefficient control by natural inhibitors. Adv. Exp. Med. Biol. 1984; 167: 111-119

28. Hornebeck W., Soleihac J.M., Velebny V., Robert L.: On the influence of the substrate (elastin) in elastase-α1 antitrypsin interactions. Pathol. Biol. 1985; 33: 281-285

29. Hornebeck W., Moczar E., Szecsi J., Robert L.: Fatty acid peptide derivatives as model compounds to protect elastin against degradation by elastases. Biochem. Pharmacol. 1985; 34: 3315-3321

30. Bruch M., Bieth J.G.: Influence of elastin on the inhibition of leukocyte elastase by α1-proteinase inhibitor and bronchial inhibitor. Potent inhibition of elastin-bound elastase by bronchial inhibitor. Biochem. J. 1986; 238: 269-273

31. Kramps J.A., Morrison H.M., Burnett D., Dijkman J.H., Stockley R.A.: Determination of elastase inhibitory activity of α1-proteinase inhibitor and bronchial antileukoprotease: different results using insoluble elastin or synthetic low molecular weight substrates. Scand. J. Clin. Lab. Invest. 1987; 47: 405-410

32. Morrison H.M., Welgus H.G., Stockley R.A., Burnett D., Campbell E.J.: Inhibition of human leukocyte elastase bound to elastin: relative ineffectiveness and two mechanisms of inhibitory activity. Amer. J. Respir. Cell. Molec. Biol. 1990; 2: 263-269

33. Cha S.: Tight-binding inhibitors - I. Kinetic behavior. Biochem. Pharmacol. 1975; 24: 2177-2185. But see corrections by Cha S. Biochem. Pharmacol. 1976; 25: 1561

34. Cha S.: Tight-binding inhibitors - III. A new approach for the determination of competition between tight-binding inhibitors and substrates. Inhibition of adenosine deaminase by coformycin. Biochem. Pharmacol. 1976; 25: 2695-2702

35. Cha S.: Tight-binding inhibitors - VII. Extended interpretation of the rate equation. Experimental designs and statistical methods. Biochem. Pharmacol. 1980; 29: 1779-1789

36. Morrison J.F.: The slow-binding and slow, tight-binding inhibition of enzyme-catalysed reactions. Trends Biochem. Sci. 1982; 7; 102-105

37. Morrison J.F., Stone S.R.: Approaches to the study and analysis of the inhibition of enzymes by slow and tight-binding inhibitors. Comments Mol. Cell. Biophys. 1985; 2: 347-368

38. Morrison J.F., Walsh C.T.: The behavior and significance of slow-binding inhibitors. Adv. Enzymol. Relat. Areas Mol. Biol. 1988; 61: 201-301

39. Bieth J.G.: Pathophysiological interpretation of kinetic constants of protease inhibitors. Bull. Eur. Physiopathol. Respir. 1980;16 (Suppl.): 183-195

40. Bieth J.G.: In vivo significance of kinetic constants of protein proteinase inhibitors. Biochem. Med. 1984; 32: 387-397

41. Baici A.: Interaction of human leukocyte elastase with soluble and insoluble protein substrates. A practical kinetic approach. Biochim. Biophys. Acta 1990; 1040: 355-364

42. Baici A.: The specific velocity plot. A graphical method for determining inhibition parameters for both linear and hyperbolic enzyme inhibitors. Eur. J. Biochem. 1981; 119: 9-14

43. Baici A., Gyger-Marazzi M.: The slow, tight-binding inhibition of cathepsin B by leupeptin. A hysteretic effect. Eur. J. Biochem. 1982; 129: 33-41

44. Baici A.: Pre-steady-state kinetic analysis of the interaction of proteinases with slow-binding inhibitors. Symp. Biol. Hung. 1984; 25: 355-367
45. Baici A., Seemüller U.: Kinetics of the inhibition of human leucocyte elastase by eglin from the leech Hirudo medicinalis. Biochem. J. 1984; 218: 829-833
46. Baici A.: Hysteretic enzyme response induced by inhibitory antibodies against human leukocyte elastase. Biol. Chem. Hoppe-Seyler 1986; 367: 245-258
47. Baici A., Pelloso R., Hörler D.: The kinetic mechanism of inhibition of human leukocyte elastase by MR889, a new cyclic thiolic compound. Biochem. Pharmacol. 1990; 39: 919-924
48. Cornish-Bowden A.: *Fundamentals of enzyme kinetics*. London, Butterworths 1979; p. 84
49. Orsi B.A., Tipton K.F.: Kinetic analysis of progress curves. Meth. Enzymol. 1979; 63: 159-183
50. Segel I.H.: *Enzyme kinetics*. New York, Wiley 1975; p. 54-64

8. Proteinase Inhibitor Candidates for Therapy of Enzyme-Inhibitor Imbalances

H. Fritz[1], J. Collins[2], M. Jochum[1]

1. Department of Clinical Chemistry and Clinical Biochemistry, University of Munich, Germany
2. Gesellschaft für Biotekhnologische Forschung, Hannover, Germany

Rationale for Therapeutic Use of Proteinase Inhibitors

Proteinase inhibitors are important regulators of proteolytic processes in the healthy organism as well as potent protectors against destructive proteolysis in various diseases.

Our knowledge on their functional role arises either from congenital or acquired deficiencies in endogenous proteinase inhibitor proteins.

Best known examples are the predisposition for lung emphysema formation because of inherited α_1PI deficiency[1] and severe bleeding disorders (disseminated intravascular coagulation, DIC) due to massive consumption of antithrombin III (AT III) e.g. in septic shock.[2]

The major target enzyme of α_1PI is the lysosomal elastase from neutrophils. This digestive proteinase is thought to play a crucial role in degradation of lung elastin fibres and thus emphysema development if it is released extracellularly from accumulating neutrophils into lung tissue over many years without being effectively inhibited due to lack of α_1PI.

Severe deficiency in α_1PI can be caused also by inflammatory events, at least locally.

For example, strong stimulation of phagocytes accumulating (e.g. neutrophils, monocytes) or present (e.g. macrophages) in an infectious or traumatic focus leads to generation and extracellular release of numerous oxidants and lysosomal digestive enzymes. α_1PI is rapidly inactivated by both types of substances, e.g. oxidants like hydrogen peroxide together with myeloperoxidase[3] as well as cysteine and metalloproteinases.[4,5]

102

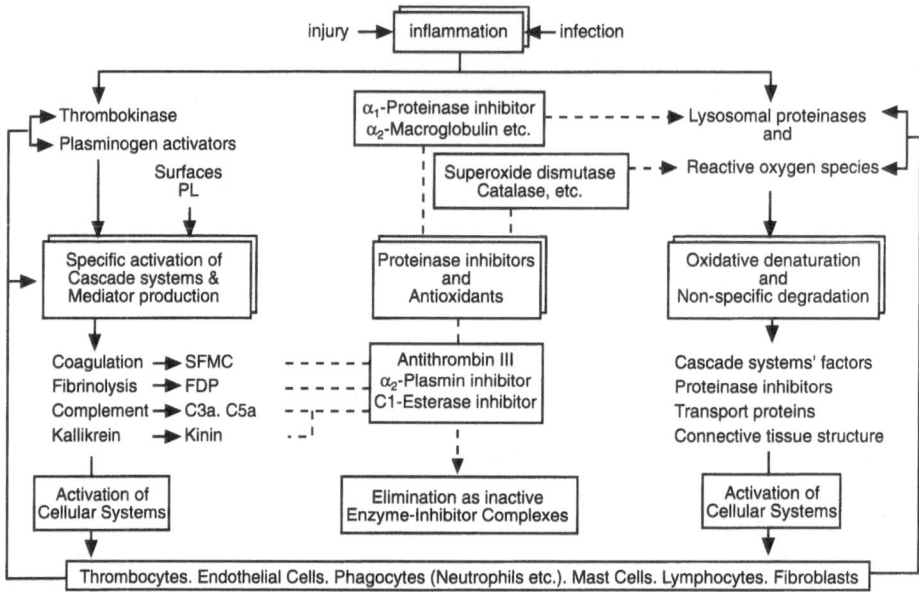

Fig. 1. Proteolysis-induced pathomechanisms in inflammatory processes. Activation of proteinase cascade systems and liberation of lysosomal proteinases concomitantly with reactive oxygen species may cause massive consumption of proteinase inhibitors which protect the organism against excessive system-specific proteolysis (see cascade systems) and unspecific proteolytic degradation by lysosomal enzymes.

As neutrophil elastase has its proteolysis optimum at slightly alkaline, i.e. physiological pH, this enzyme degrades effectively structural elements such as basal membranes, elastin and collagen fibres, fibronectin and proteoglycans as well as all kinds of humoral protein factors including the proteinase inhibitors regulating the plasmatic enzyme cascade systems (clotting, fibrinolysis, complement)[6] in the absence of inhibitors (Fig. 1).

In this way a local inflammatory process with an isolated impaired organ function due to e.g. oedema formation may become a generalized systemic inflammation leading finally to multiple organ failure and even death.[6]

Excessive consumption or destruction of proteinase inhibitors and especially of α_1PI during an inflammatory process is, therefore, a most critical event enabling the propagation of the manifold pathomechanisms inducible by proteinases which are "out of control" by their natural antagonists, the proteinase inhibitor proteins. As the natural sources for their preparation from human material are very restricted, the design of highly effective inhibitory proteins on the basis of human proteinase inhibitor molecules by molecular modelling and their production by recombinant DNA technology is the most promising approach at present to get the quantities necessary for proteinase inhibition therapy in future.[7]

Inhibitor Candidates

General comments and overview

Numerous efforts to design synthetic proteinase inhibitors (including such for neutrophil elastase) for therapeutic purposes have been not very successful so far.[8,9] The major problems such synthetic compounds are concerned with are:

1. sufficient restriction of the inhibitory specificity to avoid undesired side effects;
2. rapid elimination from all compartments of the organism.

Proteinase inhibition therapy suitable for a wider medical application has to be oriented primarily on the "functional design" of the natural endogenous inhibitors and, in special cases, on the biochemical conditions of the disease state. For example: Inhibitors designed to interfere with proteinases of the humoral cascade systems (clotting and kallikrein-kinin pathway, fibrinolysis, complement) should react either highly specifically with a certain proteinase - many of them with different functions are closely related - or resemble in their inhibition spectrum the endogenous serpins, which presumably have adopted "ideal" properties during evolution. Inhibitors designed to block lysosomal digestive proteinases (if released extracellularly) should not interfere at all with the intracellular protein breakdown, i.e. the elimination function of the phagocytes (reticuloendothelial system).

Hence, they should not be taken up into phagolysosomes or, if this occurs, they must be sensitive to oxidative and/or proteolytic inactivation in the digestive vacuole. The same holds true if such inhibitors are used for long term therapy to enable their proper inactivation, especially if their target proteinase should be, in addition to intracellular processes, involved also in an extracellular function, e.g. in penetration of phagocytes through glycoprotein membrane layers.[10,11]

On the other hand, under severe inflammatory conditions (e.g. multiple injuries, septicaemia, isolated or multiple organ failure like ARDS or MOF) phagocytes may produce high amounts of oxygen free radicals, hydrogen peroxide etc. and even discharge their lysosomal contents;[6] in such dramatic pathological events oxidation-resistant proteinase inhibitors might be much more effective as "anti-inflammatory drugs".[7] Further, to minimize the risk of a response of the immune system, clinically administered inhibitors should resemble as closely as possible the endogenous inhibitor proteins. Inhibitor candidates which according to our opinion are most suitable for proteinase inhibition therapy as discussed above or elsewhere[7] are listed in Table I.

α_1 Proteinase Inhibitor (α_1PI)

The predominant natural antagonist of neutrophil elastase, which is already

Table I. Inhibitor candidates of human origin suitable for proteinase (neutrophil elastase etc.[a]) inhibition therapy

α_1 PI	native form	- glycoprotein	-oxidizable
	r-variants	- proteins	-oxidation resistant
MPI	native form	- miniprotein	-oxidizable
	r-variants	- miniproteins	- oxidation resistant

Aprotinin (miniprotein) homologous domains[b] (no.) in:
1. Inter-α-trypsin inhibitor complex as Bikunin (2)
2. Alzheimer amyloid protein precursor Pre A4 (1)
3. Lipoprotein-associated coagulation inhibitor (3)

Kazal-type inhibitors (miniproteins): r-variants of:
1. Pancreatic secretory trypsin inhibitor, PSTI
2. Seminal acrosin-trypsin inhibitor, HUSI-II

[a] and/or cathepsin G and/or mast cell chymase; [b] recombinant variants; r = recombinant

therapeutically given to α_1PI-deficient individuals with lung emphysema, can be isolated from normal human blood only in limited quantities.

Therefore, various possibilities for its production by genetic engineering are presently under investigation.[12,13]

The same is true for an oxidation resistant artificial mutant of α_1PI, Val[358] α_1PI,[14,15] as well as for a naturally occurring mutant (and variants thereof), Arg[358] α_1PI, which is a strong inhibitor of thrombin and plasma kallikrein[15,16] (Table II). However, as long as these α_1PI homologues cannot be produced in sufficient

Table II. Residues in the reactive site region and inhibitory specificity of serpins

Serpin or variant	Major target enzyme	Residues in positions					
		P_2	P_1	P'_1	P'_2	P'_3	P'_4
α_1 PI (α_1 AT)	neutrophil E	Pro	**Met**	Ser	Ile	Pro	Pro
r-variant*	neutrophil E	Pro	**Val**	Ser	Ile	Pro	Pro
α_1 PI-Pittsburgh	thrombin	Pro	**Arg**	Ser	Ile	Pro	Pro
Antithrombin III	thrombin	Gly	**Arg**	Ser	Leu	Asn	Pro

AT = antitrypsin; E = elastase; *oxidation resistant, cf. Table I quantities in the natural glycoprotein form, their suitability in pharmacological terms (distribution within the organism, elimination rate, etc.) and a possible immune response against the carbohydrate chain free, "naked" proteins has to be carefully considered.

Mucus proteinase inhibitor (MPI)

The mucus proteinase inhibitor MPI, also known as antileucoprotease (ALP) or secretory leucocyte proteinase inhibitor (SLPI), is the predominant natural antagonist of neutrophil elastase in all mucous secretions of the organism.[17] It represents together with α_1PI the main antiprotease shield of the upper airways (molar ratio of MPI=α_1PI > 1) and the lung (MPI=α_1PI < 1).[7]

Despite the presence of 8 disulphide bridges within the molecule (Fig. 2), the

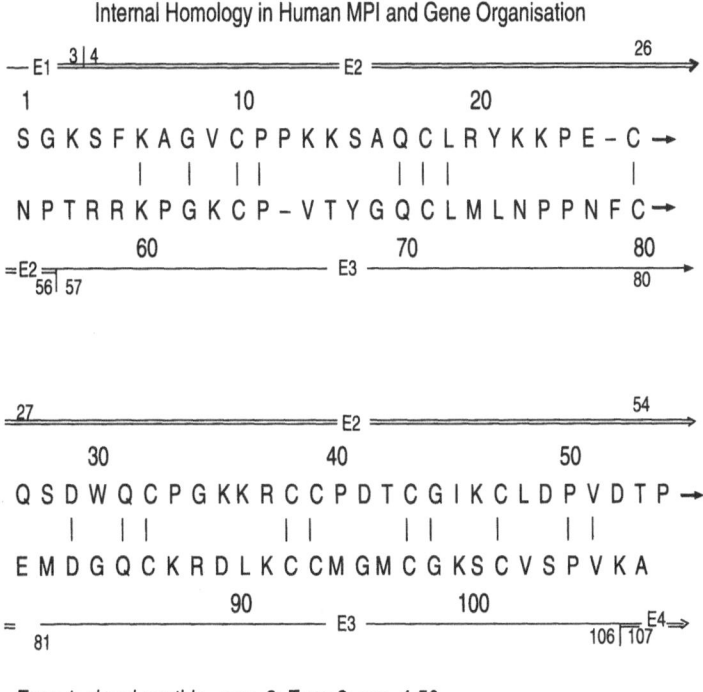

Exon 1: signal peptide - pos. 3; Exon 2: pos. 4-56
Exon 3: pos. 57-106; Exon 4: pos. 107 + poly A recogn. signal

Fig. 2. Primary structure, internal sequence homology and genomic organization of the human mucus proteinase inhibitor MPI.
The exon-intron organization reflects the two structural homologous domains of the MPI molecule. Present evidence suggests that it is the C-terminal domain which is inhibitorily active against neutrophil elastase and cathepsin G as well as chymotrypsin and trypsin (P$_1$= Leu[72], P$_1$'= Met[73])

natural mature form of the miniprotein MPI can be produced already in sufficient quantity and purity to start investigation of its therapeutic effectiveness in animal models[18] and in patients suffering from emphysema and cystic fibrosis (Synergen, Boulder/Colorado, USA). For first therapeutic approaches to elucidate its potential anti-inflammatory effectiveness, an oxidation resistant MPI variant is also available (Grünenthal GmbH, Stolberg/Rheinland, Germany). In this variant the four Met residues present in the second domain, which is responsible for neutrophil elastase and cathepsin G inhibition (Fig. 2), have been exchanged by aliphatic amino acid residues without impairment of the inhibitory activity.[19]

Aprotinin homologues

The miniprotein aprotinin from bovine mast cells, a proteinase inhibitor with rather low specificity, has been used in medical therapy uncritically for a long time to treat numerous diseases[20] before clinically clearly effective dosages were applied more recently.[21,22]

Proteinase inhibition therapy with higher dosages of aprotinin proved to be especially valuable in open heart surgery with extracorporeal circulation whereby blood loss and transfusion requirement could be highly significantly reduced most probably due to effective plasmin inhibition.[23,24]

Polypeptide domains which are structurally closely related to bovine aprotinin have been found to occur also in human high molecular mass protein complexes or proteins (Table I). The "Bikunin" molecule present in the protein complex of the inter-alpha-trypsin inhibitor consists essentially of two aprotinin-like domains, the N-terminal one (D_1) being responsible for inhibition of neutrophil elastase and the C-terminal domain (D_2) for trypsin inhibition.[25,26] Exchange of the two Met residues in the reactive site region of the human Bikunin domain D_1 by Leu residues in the bovine molecule (Table III) leads to a dramatic increase in the affinity to neutrophil elastase and cathepsin G.[26] Hence, artificial mutants of the human Bikunin molecule with high specificity and strong affinity to a certain target proteinase like the neutrophil elastase may be designed - e.g. by comparison with naturally occurring mutants or by molecular modelling of the structure of the inhibitor in its complex with the proteinase - and finally produced by recombinant DNA techniques.[27] A similar approach should be possible with the aprotinin-like domain(s) present in Alzheimer amyloid protein precursor PreA4,[28] and lipo-protein-associated coagulation inhibitor.[29]

Kazal-type Inhibitors

Kazal-type inhibitors comprise a family of miniproteins (single inhibitory domain) or proteins (composed of several such domains: multiheaded) with primary structures similar to the sequence of bovine pancreatic secretory trypsin

Table III. Reactive site residues in the N-terminal domains (D_1) responsible for chymotrypsin and neutrophil elastase inhibition of the human and bovine Bikunin molecules. The C-terminal domains (D_2) responsible for trypsin inhibition have identical sequences in this region for both species

Domain	Subsite positions					
	P_3	P_2	P_1	P'_1	P'_2	P'_3
human D_1	Pro	Cys	**Met**	Gly	**Met**	Thr
bovine D_1	Pro	Cys	**Leu**	Gly	**Leu**	Phe
D_2	Pro	Cys	**Arg**	Ala	Phe	Ile

inhibitor first described by Kazal;[30] they are widely distributed in vertebrates.[31,32] In the human organism two Kazal-type inhibitors, each of them single-headed, have been identified unequivocally by amino acid sequence analysis. The first was pancreatic secretory trypsin inhibitor, PSTI,[33] and more recently the trypsin-acrosin inhibitor HUSI-II (human seminal acrosin inhibitor II) which occurs in male genital tract organs and secretions.[34,35]

Due to the extensive studies of M. Laskowski et al.[31,32] on structure-function relationships of Kazal-type avian ovomucoid domains, a large data base for predictable alterations of the inhibitory affinity and specificity by suitable substitutions of a single or a few residues in the reactive peptide sequence is available. In view of this knowledge the human PSTI which inhibits strongly its natural antagonist, the pancreatic trypsin, was chosen for a protein design project.[36] Our aim was to develop inhibitors with high affinity against human neutrophil

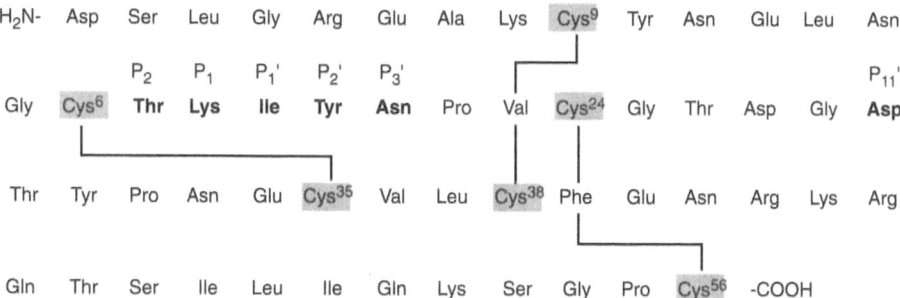

Fig. 3. Primary structure of the human pancreatic secretory trypsin inhibitor, PSTI, with subsite positions (residues) in most intimate contact to the enzyme(s) in the complex and disulphide bridges as indicated.

Table IV. Inhibitory specificity of PSTI variants expected on the basis of M. Laskowski's structure-function algorithm of Kazal-type ovomucoid inhibitors

PSTI variant	Subsite position						Specificity for
	P_2	P_1	P'_1	P'_2	P'_3	P'_{11}	
PSTI n	Thr	Lys	Ile	Tyr	Asp	*Asn*	T
PSTI 0	Thr	Lys	Ile	Tyr	Asn	Asp	T
PSTI 1	Thr	**Leu**	Ile	Tyr	Asn	Asp	E and Ch
PSTI 2	Thr	**Leu**	Ile	Tyr	Asp	*Asn*	E and Ch
PSTI 3	Thr	**Tyr**	*Glu*	Tyr	*Arg*	Asp	E
PSTI 4	Thr	**Leu**	*Glu*	Tyr	*Arg*	Asp	E and Ch
PSTI 5	Thr	**Val**	*Glu*	Tyr	*Arg*	Asp	E
PSTI 6S	Thr	**Leu**	*Glu*	Tyr	Asn	Asp	E and Ch
PSTI 7	Thr	**Leu**	Ile	Tyr	*Arg*	Asp	E and Ch
PSTI 8	Thr	**Val**	*Glu*	*Leu*	Asn	Asp	E
PSTI 9	Thr	**Val**	*Glu*	*Leu*	*Arg*	Asp	E
PSTI 10	*Pro*	**Lys**	Ile	Tyr	Asp	*Asn*	T
PSTI 11	*Pro*	**Leu**	*Glu*	Tyr	*Arg*	Asp	E and Ch
PSTI 12	*Pro*	**Val**	*Glu*	Tyr	*Arg*	Asp	E
PSTI 13	Thr	**Ile**	*Glu*	Tyr	Asn	Asp	E?
PSTI 14	Thr	**Arg**	*Glu*	Tyr	Asn	Asp	T
PSTI 15	Thr	**Phe**	*Glu*	Tyr	Asn	Asp	Ch and C-G
PSTI 16	Thr	**Ala**	*Glu*	Tyr	Asn	Asp	E
PSTI 17	Thr	**Val**	Ile	Tyr	Asn	Asp	E
PSTI 18	Thr	**Ile**	Ile	Tyr	Asn	Asp	E?
PSTI 19	Thr	**Val**	Ile	Tyr	Asp	*Asn*	E

n = natural native inhibitor; C-G = cathepsin G; Ch = chymotrypsin; E = elastase; T = trypsin

elastase and cathepsin G with PSTI as model compound.[37,38]

The primary structure of PSTI and its subsite positions (residues) in most intimate contact with the target enzyme(s) in the complex are shown in figure 3. The primary inhibitory specificity of various artificial mutants of PSTI produced by recombinant DNA techniques is indicated in table IV.

In Table V, PSTI variants exhibiting highest affinity for the chosen target enzymes are listed together with the K_i values which reflect the influence of amino acid exchanges in certain subsite positions of the PSTI molecule on the affinity. Even further improvement of specificity and selectivity turned out to be possible by additional substitutions in position 21 or 36.[37]

At present, similar studies are being performed with the trypsin-acrosin inhibitor HUSI-II as model compound in other laboratories.

Table V. Inhibition of chymotrypsin (Ch) and neutrophil elastase (E) by r-PSTI variants

The effect of the P_1 and P'_1 residues (all with P'_3 = Arg)

PSTI variant	P_1	P'_1	$K_i(Ch)$	$K_i(E)$
PSTI-3	Tyr	Glu	1.6×10^{-11}	$> 10^{-7}$
PSTI-5	Val	Glu	3.1×10^{-7}	1.5×10^{-11}
PSTI-4	Leu	Glu	2.4×10^{-11}	3.7×10^{-11}
PSTI-7	Leu	Ile	8.0×10^{-9}	2.5×10^{-11}

The effect of the P'_3 residue (P_1 = Leu; P'_{11} = Asp)

PSTI variant	P'_3	P'_1	$K_i(Ch)$	$K_i(E)$
PSTI-1	Asn	Ile	5.0×10^{-8}	5.0×10^{-11}
PSTI-7	Arg	Ile	8.0×10^{-9}	2.5×10^{-11}
PSTI-6	Asn	Glu	2.0×10^{-8}	2.5×10^{-10}
PSTI-4	Arg	Glu	2.4×10^{-11}	3.7×10^{-11}

K_i = dissociation equilibrium constant of the enzyme-inhibitor complex in mol/l

Table VI. Natural proteinase inhibitors in therapy

Indication	Applied	Use envisaged	Target enzymes
hyperfibrinolysis, shock states	bovine aprotinin		plasmin, plasma and tissue kallikrein
coagulopathy, DIC	antithrombin III	r-hirudin	thrombin
angioneurotic oedema	C1 inhibitor		plasma kallikrein, F XIIa, C1 esterase
emphysema	α_1PI		
inflammation (sepsis, ARDS, MOV, etc.)		r-α_1PI, r-eglin, r-MPI (ALP, SLPI)*	neutrophil elastase and cathepsin G

α_1PI = α_1 proteinase inhibitor; r = recombinant
*mucus proteinase inhibitor (ALP = antileucoprotease, SLPI = secretory leucocyte proteinase inhibitor)

Conclusion

The given data show clearly that miniprotein inhibitors with highest affinity and selectivity for certain proteinases including further desired properties (e.g. oxidation resistance) can be prepared. Such artificial mutants of natural regulators of proteinases are of great value for both biochemical investigations of structure-function relationships and therapeutic experimental and clinical studies. Hence, the modern methods of molecular modelling and biotechnology have provided us with suitable techniques to enable the design and production of inhibitors for effective proteinase inhibition therapy in the near future (Table VI). The major problem, however, which still has to be solved in this respect is the preparation of kilogram amounts of inhibitory drugs in highest purity for more extensive experimental and clinical studies. A successful outcome of these studies would imply the production of such inhibitors in sufficient amounts for common medical use.

References

1. Crystal R.G.: α_1-Antitrypsin deficiency, emphysema, and liver disease. J. Clin. Invest. 1990; 85: 1343-1352
2. Büller H.R., Ten Cate J.W.: Acquired antithrombin III deficiency: laboratory diagnosis, incidence, clinical implications, and treatment with antithrombin III concentrate. Am. J. Med. 1989; 87 (Suppl. 3B): 44-48
3. Ossanna P.J., Test S.T., Matheson N.R., Regiani S., Weiss St.J.: Oxidative regulation of neutrophil elastase-alpha$_1$-proteinase inhibitor interactions. J. Clin. Invest. 1986; 77: 1939-1951
4. Johnson D.A., Barrett A.J., Mason R.W.: Cathepsin L inactivates α_1-proteinase inhibitor by cleavage in the reactive site region. J. Biol. Chem. 1986; 261: 14748-14752
5. Desrochers P.E., Weiss St.J.: Proteolytic inactivation of α_1-proteinase inhibitor by a neutrophil metalloproteinase. J. Clin. Invest. 1988; 81: 1646-1650
6. Jochum M., Fritz H.: Pathobiochemical mechanisms in inflammation. In: Faist E., Ninnemann J.L., Green D.R.(Eds.): *Immune Consequences of Trauma, Shock and Sepsis*. Berlin-Heidelberg, Springer Verlag, 1989; 165-172
7. Jochum M., Fritz H.: Elastase and its inhibitors in intensive care medicine. Biomed. Progress 1990; 3: 55-59
8. Groutas W.C.: Inhibitors of leukocyte elastase and leukocyte cathepsin G. Agents for the treatment of emphysema and related ailments. Med. Res. Rev. 1987; 7: 227-241
9. Sandler M., Smith H.J.(Eds.): *Design of Enzyme Inhibitors as Drugs*. Oxford, New York, Tokyo; Oxford University Press 1989
10. McGowan S.E., Murray J.J.: Direct effects of neutrophil oxidants on elastase-induced extracellular matrix proteolysis. Am. Rev. Respir. Dis. 1987; 135: 1286-1293
11. Travis J., Fritz H.: Potential problems in designing elastase inhibitors for therapy. Am. Rev. Respir. Dis. 1991; 143: 1412-1415
12. Hubbard R.C., McElvaney N.G., Sellers S.E., Healy J.T, Czerski D.B., Crystal R.G.: Recombinant DNA-produced alpha$_1$-antitrypsin administered by aerosol augments lower respiratory tract antineutrophil elastase defenses in individuals with α_1-antitrypsin deficiency. J. Clin. Invest. 1989; 84: 1349-1354

13. Gilardi P., Courtney M., Pavirani A., Perricaudet M.: Expression of human α_1-antitrypsin using a recombinant adenovirus vector. FEBS Lett. 1990; 267: 60-62

14. George P.M., Vissors M.C.M., Travis J., Winterbourn C.C., Carrell R.W.: A genetically engineered mutant of α_1-antitrypsin protects connective tissue from neutrophil damage and may be useful in lung disease. Lancet 1984; 1426-1428

15. Courtney M., Jallat S., Tessier L.H., Crystal R., Lecocq J.P.: The construction of novel protease inhibitors by modification of the active centre of α_1-antitrypsin. Phil. Trans. R. Soc. Lond. 1986; A317: 381-390

16. Schapira M., Ramus M.A., Jallat S., Carvalho D., Courtney M.: Recombinant α_1-antitrypsin Pittsburg (Met^{358} Arg) is a potent inhibitor of plasma kallikrein and activated factor XII fragment. J. Clin. Invest. 1985; 76: 635-637

17. Fritz H.: Human mucus proteinase inhibitor (human MPI). Human seminal inhibitor I (HUSI-I), antileukoprotease (ALP), secretory leukocyte protease inhibitor (SLPI). Biol. Chem. Hoppe-Seyler 1988; 369 (Suppl.): 79-82

18. Lucey E.C., Stone Ph.J., Ciccolella D.E., Breuer R., Christensen T.G., Thompson R.C., Snider G.L.: Recombinant human secretory leukocyte-protease inhibitor: In vitro properties and amelioration of human neutrophil elastase-induced emphysema and secretory cell metaplasia in the hamster. J. Lab. Clin. Med. 1990; 115: 224-232

19. Heinzel-Wieland R., Ammann J., Steffens G.J., Flohe L.: *Neue Serinprotease-Inhibitor-Proteine, diese enthaltende Arzneimittel und DNA-Sequenzen, die für diese Proteine codieren und Verfahren zur Herstellung dieser Proteine, Arzneimittel und DNA-Sequenzen.* Patentschrift Nr.DE 3841873 A1. 1990

20. Fritz H., Wunderer G.: Biochemistry and applications of aprotinin, the kallikrein inhibitor from bovine organs. Arzneimittel-Forsch./Drug. Res. 1983; 33: 479-494

21. Jochum M., Müller-Esterl W.: Bestimmung von Aprotinin-Plasmakonzentrationen nach therapeutischer Anwendung von Trasylol. In: Dudziak R., Kirchhoff P.G., Reuter H.D., Schumann F. (Hrsg.). *Proteolyse und Proteinaseninhibition in der Herz-und Gefäßchirurgie.* Stuttgart-New York, Schattauer-Verlag, 1985; p. 157-167

22. Clasen C., Jochum M., Müller-Esterl W.: Feasibility study of very high aprotinin dosage in polytrauma patients. In: Schlag G., Redl H. (Eds.): *Progress in Clinical and Biological Research. Subseries: First Vienna Shock Forum. Part A: I. Pathophysiological Role of Mediators and Mediator Inhibitors in Shock.* New York, A.R. Liss Inc. 1987; 175-183

23. Bidstrup B.P., Royston D., Taylor K.M.: Reduction in blood loss and blood use after cardiopulmonary bypass with high dose aprotinin (Trasylol). J. Thorac. Cardiovasc. Surg. 1989; 97: 364-372

24. Dietrich W., Spannagl M., Jochum M., Wendt P., Schramm W., Baranky A., Sebening F.: Influence of high-dose aprotinin treatment on blood loss and coagulation patterns in open-heart surgery. Anesthesiology, 1990;73:1119-1126

25. Gebhard W., Leysath G., Schreitmüller T.: Inter-α-trypsin inhibitor is a complex of three different protein species. Biol. Chem. Hoppe-Seyler 1988; 369 (Suppl.): 19-22

26. Gebhard W., Hochstrasser K.: Inter-α-trypsin inhibitor and its close relatives. In: Barrett A.J., Salvesen G. (Eds.): *Proteinase Inhibitors. Research monographs in cell and tissue physiology,* Vol. 12. Amsterdam, Elsevier, 1986; p. 389-401

27. Gebhard W.: *Inter-α-Trypsininhibitor und Carboxypeptidase N. Aufklärung der Primärstrukturen und Konstruktion varianter Proteine.* Habilitation thesis. The Ludwig-Maximilians-University of Munich, 1988

28. Oltersdorf T., Fritz L.C., Schenk D.B., Lieberburg I., Johnson-Wood K.L., Beattie E.C., Ward P.J., Blacher R.W., Dovey H.F., Sinha S.: The secreted form of the Alzheimer's amyloid precursor

protein with the Kunitz domain is protease nexin II. Nature 1989; 341: 144-147

29. Wun T.Ch., Kretzmer K.K., Girard T.J., Miletich J.P., Broze G.J.: Cloning and characterization of a cDNA coding for the lipoprotein-associated coagulation inhibitor shows that it consists of three tandem Kunitz-type inhibitory domains. J. Biol. Chem. 1988; 263: 6001-6004

30. Kazal L.A, Spicer D.S., Brahinsky R.A.: Isolation of a crystalline trypsin-inhibitor anticoagulant protein from pancreas. J. Am. Chem. Soc. 1948; 70: 3034-3040

31. Laskowski M.Jr., Kato I.: Protein inhibitors of proteinases. Annu. Rev. Biochem. 1980;49: 593-626

32. Laskowski M.Jr.: Protein inhibitors of serine proteinases-Mechanism and classification. In: Friedman M. (Ed.): *Nutritional and toxicological significance of enzyme inhibitors in foods.* New York, Plenum Press,1986;1-17

33. Bartelt D.C., Shapanka R., Greene L.J.: The primary structure of the human pancreatic trypsin inhibitor. Arch. Biochem. Biophys. 1977; 179: 189-199

34. Fink E., Hehlein-Fink C., Eulitz M.: Amino acid sequence elucidation of human acrosin-trypsin inhibitor (HUSI-II) reveals that Kazal-type proteinase inhibitors are structurally related to β-subunits of glycoprotein hormones. FEBS Lett. 1990; 270: 222-224

35. Schiessler H., Arnhold M., Fritz H.: Characterization of two proteinase inhibitors from human seminal plasma and spermatozoa. In: Fritz H., Tschesche H., Greene L.J., Truscheit E.: *Proteinase Inhibitors.* Berlin-Heidelberg-New York, Springer-Verlag 1974; p. 147-155

36. Maywald F., Böldicke T., Gross G., Frank R., Blöcker H., Meyerhans A., Schwellnus K., Ebbers J., Bruns W., Reinhardt G., Schnabel E., Schröder W., Fritz H., Collins J.: Human pancreatic secretory trypsin inhibitor (PSTI) produced in active form and secreted from "Escherichia coli". Gene 1988; 68: 357-369

37. Collins J., Szardenings M., Maywald F., Fritz H., Bruns W., Reinhardt G., Schnabel E., Schröder W., Blöcker H., Reichelt J., Schomburg D.: Design of efficient human leukocyte elastase inhibitors:.Variants of human pancreatic secretory trypsin inhibitor (hPSTI). In: Blöcker H., Collins J., Schmid R.D., Schomburg D. (Eds.): *Advances in Protein Design International Workshop 1988* (GBF Monographs, Vol. 12). Weinheim, VCH Verlagsgesellschaft, 1989; p. 201-210

38. Collins J., Szardenings M., Maywald F., Blöcker H., Frank R., Hecht H.J., Vasel B., Schomburg D., Fink E., Fritz H.: Human leukocyte elastase inhibitors: Designed variants of human pancreatic secretory trypsin inhibitor (hPSTI). Biol. Chem. Hoppe-Seyler 1990; 371 (Suppl.): 29-36

9. Antileucoprotease (Secretory Leucocyte Proteinase Inhibitor), a Major Proteinase Inhibitor in the Human Lung

J. A. KRAMPS, J. STOLK, A. RUDOLPHUS, J. H. DIJKMAN
Department of Pulmonology, University Hospital of Leiden, The Netherlands

Introduction

The current most widely accepted explanation for the destruction of alveolar walls, as seen in pulmonary emphysema, is an unrestrained elastolytic activity in the peripheral lung due to a local imbalance between neutrophil elastase and the antineutrophil elastase screen.[1-3] This hypothesis has arisen mainly from the observation that heritable severe deficiency of α_1-proteinase inhibitor (α_1PI) is associated with development of the disease at a relatively young age. α_1PI, present in alveolar airspaces as a transudated protein from the circulation, plays an important role in the regulation of neutrophil elastase activity.

Cigarette smoking, the primary environmental risk factor associated with emphysema[4,5], also may alter the elastase/elastase inhibitor balance in the lung.[1-3] The contribution of smoking in disturbing the balance is explained by a smoke-induced recruitment of inflammatory cells releasing their lysosomal enzymes, resulting in a local increase of elastase burden. In addition, reactive oxygen species, present in smoke and produced by stimulated inflammatory cells, are capable of inactivating α_1PI.

However, irrespective of a large body of circumstantial evidence, there is still no definite proof for a disturbance of the elastase/elastase inhibitor balance in the lungs of smokers as the cause of alveolar wall destruction.

Though a severe deficiency of α_1PI and cigarette smoking are the two major risk factors for the development of emphysema, it has become clear that these factors are not invariably linked to the disease.[3,6,7] The presence and severity of emphysema may vary considerably among smokers and among α_1PI deficient individuals.

114

Consequently, in all probability multiple factors are involved in the pathogenic mechanism leading to alveolar destruction. Such a factor, which may play a modulating role in the development of emphysema, is antileucoprotease[8], also called secretory leucocyte proteinase inhibitor.[9]

This report will focus on this proteinase inhibitor which was first described as present in airway secretions by Hochstrasser et al. in 1972.[10]

Properties of Antileucoprotease

Antileucoprotease (ALP), which can be isolated from acid-treated bronchial mucus by trypsin or chymotrypsin affinity chromatography,[11,12] is a low molecular weight non-glycosylated highly basic protein (pI>9) of 12 kDa. ALP is able to inhibit several serine proteinases, including trypsin and α-chymotrypsin from bovine pancreas,[12] mast cell chymase[13] and the neutrophil lysosomal enzymes elastase and cathepsin G.[12] Based on kinetic measurements[14] it was concluded that ALP acts *in vivo* as an inhibitor of neutrophil elastase with which a highly stable complex can be formed very rapidly ($k_{ass} = 6.4 \times 10^6 M^{-1}s^{-1}$; $k_{diss} = 2.3 \times 10^{-3}s^{-1}$).

In 1986 the complete amino-acid sequence of ALP was elucidated by two independent laboratories[9,15] and shortly after that, the crystal structure of ALP in its complex with α-chymotrypsin was solved by Grütter et al.[16] The protein of 107 amino acids comprises two consecutive highly homologous domains of approximately 50 amino acids, each containing 4 disulphide bridges (Fig. 1). A correlation exists between the two domains of the ALP molecule and the organization of its gene on chromosome 16, that is to say the coding information of the two domains corresponds with two separate exons.[17]

The X-ray crystallographic analysis revealed that the two domains of ALP have a similar architecture and are spatially well separated.[16]

Based on a slight homology with the active site of Kazal type inhibitors and on

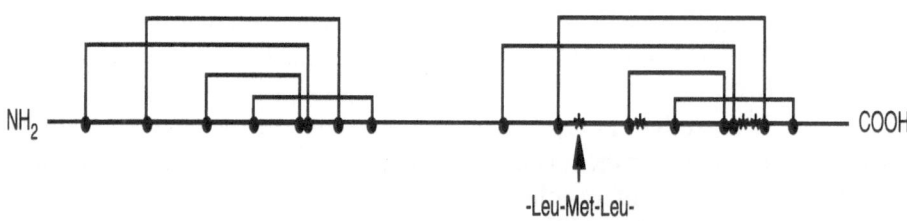

Fig. 1. Presentation of the two domain structure of antileucoprotease showing the disulphide bridge arrangement within each domain. Half-cystine residues are indicated by black dots and the 4 methionine residues in the ALP molecule by crosses. The proteinase binding site in the second COOH-terminal domain is also shown (arrow).

the localization where limited proteolytic cleavage of ALP has been observed[9,15], it was proposed that the antitrypsin activity is represented by Leu 19-Arg 20-Tyr 21 in the first NH_2-terminal domain, whereas the antielastase and antichymotrypsin activities are represented by Leu 72-Met 73-Leu 74 in the COOH-terminal domain. By applying X-ray crystallographic techniques[16], the exact position of the binding site of ALP with chymotrypsin indeed has been defined as Leu 72-Met 73-Leu 74 (Fig. 1). Preliminary experiments performed in the same laboratory showed that neutrophil elastase also fits to this binding site in the second domain.

Concerning the proposed trypsin inhibitory site in the NH_2-terminal domain of ALP, we recently obtained evidence that this prediction is not correct.[18,19] Treatment of ALP with different amounts of methionine-selective reagents like myeloperoxidase-generated reactive oxygen species, N-chlorosuccinimide or cisplatinum(II)diammine dichloride, resulted in a dose-dependent inactivation of all inhibitory activities. This strongly suggests that methionine residues are involved in inhibitory activity against trypsin, chymotrypsin and elastase.

Besides Met73, ALP contains three additional methionine residues all located in the second COOH-terminal domain (Fig. 1). Furthermore, localization of the trypsin inhibitory site in the second domain, close to or at the position of the elastase inhibitory site, was supported by the observation that elastase is able to displace trypsin from the inhibitor molecule.[19] In all probability, ALP would bind elastase and trypsin simultaneously in case the inhibitory sites of the two enzymes are located at different domains as initially was supposed.

Very recently, Eisenberg et al.[20] performed experiments, using ALP mutants prepared by site direct mutagenesis, which confirm that antielastase, antichymotrypsin and antitrypsin sites all are located in the COOH-terminal domain of the inhibitory molecule. Consequently, at this moment it is not known what function resides in the NH_2-terminal domain of the ALP molecule.

Antileucoprotease in the Human Lung

ALP is produced by a variety of secretory cells in the human body and can be found in mucous secretions including nasal and bronchial mucus, saliva, tear fluid, cervical mucus and seminal plasma[21,22] (Table I). On the basis of its widespread distribution in these mucous secretions, it is concluded that ALP plays a role in the protection of mucous membranes of external orifices against elastase released from neutrophils during inflammatory responses. In the large airways of the human lung, where ALP is the major inhibitor of neutrophil elastase[23,24], we have shown by immunostaining methods that the inhibitor is produced by submucosal glands and bronchial nonciliated epithelial cells.[25-27]

In addition to its production at the bronchial level, we observed that ALP also is produced at the level of membranous and respiratory bronchioles by nonciliated

Table I. Quantity of antileucoprotease in human mucous secretions

	ALP/µg/ml*
Nasal secretion	150
Bronchial mucus	60
Saliva	20
Tear fluid	20
Cervical mucus	70
Seminal fluid	30
Serum/plasma	0.06

*Mean values as determined by ELISA. ALP concentrations may vary considerably between specimens of the same kind.

epithelial cells being identified as Clara cells and globet cells.[26]

This finding suggested that ALP may be a factor affecting the elastase/elastase inhibitor balance in the peripheral lung and consequently may play a role in the pathogenesis of emphysema. However, evaluating the contribution of ALP to the elastase inhibitor screen in the peripheral airspaces, it was observed that the molar concentration of ALP amounted to approximately 10% of that of α_1PI.[24,28] Notwithstanding its minor contribution to the inhibitor potency of the lining fluid of the bronchiolar epithelium, evidence has been obtained recently that ALP may exert a protective effect on elastase-mediated degradation of the extracellular matrix in lung alveolar septa. Applying an indirect immunofluorescence technique in order to search for extracellular localizations of ALP in the human lung, the inhibitor was detected in the connective tissue matrix, including that of the alveolar septa.[29] Furthermore, the results of this study strongly suggested that the extracellular ALP is present exclusively along the elastin fibres. This observation indeed has been confirmed by an immunoelectron microscopic study using an ALP-specific gold-labelling procedure.[30] Thus, ALP is present in the direct vicinity of a major target compound of neutrophil elastase, viz. elastin fibres, destruction of which is thought to be the key mechanism of emphysema.

Potency of Antileucoprotease to Inhibit Degradation of Extracellular Matrix Molecules

With regard to the presence of ALP along elastin fibres in the parenchymal matrix of the lung, a most relevant question is if ALP may function *in vivo* as a potent inhibitor of elastase-mediated matrix degradation. Bruch and Bieth[31] investigated the ability of α_1PI and ALP to inhibit elastin-bound elastase and observed significant differences between the two inhibitors. Part of the elastin-bound elastase was found to be resistant to the inhibitory action of α_1PI. In contrast, ALP inhibited the elastin-bound elastase as efficiently as it inhibited the free enzyme. Moreover, the fraction of elastin-bound elastase which remained active in the presence of excess α_1PI could be inhibited by ALP. Evidence has been found by Morrison et al.[32] for two distinct mechanisms of inhibition of elastin-bound elastase. They observed that α_1PI enhances dissociation of elastase from elastin, whereafter the enzyme is inhibited in solution. Contrary to α_1PI, ALP does not enhance dissociation of the elastase-elastin complex but is able to inhibit the enzyme while it is bound to elastin. One may speculate, as already suggested several years ago[33], that ALP functions *in vivo* mainly as an inhibitor of elastin-bound elastase, whereas α_1PI takes care of inhibiting free elastase.

In emphysema, destruction of alveolar walls may occur when polymorphonuclear leucocytes (PMN's) migrate from the capillaries via the interstitium into the alveolar space.[34] During migration, these cells may mediate destruction of interstitial matrix molecules by creating an interface between cell and matrix wherein entry of proteinase inhibitors is restricted.[35,36]

In order to mimic these events and to determine the potency of α_1PI and ALP to inhibit PMN-mediated matrix degradation, a model was used in which fibrinogen matrix degradation occurred mediated by firmly attached PMN's.[37,38] In this study, performed in collaboration with P. Davies and coworkers (Merck, Sharp, and Dohme Research Laboratories, Rahway, USA) attachment of PMN's onto a fibrinogen coated surface was facilitated by tumour necrosis factor alpha (TNF α).[37] Fibrinogen degradation, after PMN stimulation, was measured by a specific immunoassay for a neutrophil elastase generated cleavage product of the fibrinogen A-α chain [A-α(1-21)].[38] Inhibition of PMN-mediated fibrinogen degradation was tested at different doses of α_1PI or ALP.[39] In this study, half-maximum PMN-mediated fibrinogen degradation was observed at an α_1PI concentration of 220±98 nM (=IC_{50}; mean ± SD, n=7). For ALP a significant lower IC_{50} value of 85±30 nM (n=7) has been obtained (Table II). Thus, in this model ALP is a more potent regulator of matrix degradation than α_1PI. Differences in sensitivity to oxidative inactivation by reactive oxygen species (ROS), produced by stimulated cells, may explain the observed differences in potency between both inhibitors. However, experiments performed with PMN's from patients with chronic granulomatous disease (CGD), who are not capable of producing ROS, showed a similar difference in potency between ALP and α_1PI as observed in experiments performed with PMN's from normal individuals (see Table II).

By comparing for each inhibitor the IC_{50} value obtained with normal and CGD-PMN's we concluded that the two inhibitors are equally sensitive to ROS-mediated inactivation (Table II). Most probably, ALP is more potent than α_1PI to inhibit matrix degradation by firmly attached PMN's because ALP may have better access to the

Table II. Potency of ALP and α_1PI in the regulation of PMN-mediated degradation of fibrinogen

| | IC_{50} * (mean ± SD) | |
	α_1PI	ALP
normal PMN°	60 ± 5 nM	17 ± 3 nM
normal PMN after stimulation §	220 ± 98 nM	85 ± 30 nM
CGD-PMN after stimulation §	150 ± 21 nM	50 ± 13 nM

*: concentration of active inhibitor giving 50% reduction of PMN-mediated fibrinogen degradation.
°: effect of inhibitors on fibrinogen degradation by normal PMN's in the presence of TNFα.
§: effect of inhibitors of fibrinogen degradation by adherent PMN's stimulated by cytochalasin B and formyl-Met-Leu-Phe. PMN's were obtained from normal individuals or from patients with chronic granulomatous disease (CGD).

interface created between cell and matrix, due to its relatively small molecular size and/or because of its highly basic character.

The high potency of ALP to inhibit proteolysis at a neutrophil-substrate interface has very recently been confirmed by Rice and Weiss.[40] These investigators found that fibronectin or elastin could be protected from adherent neutrophils by ALP, whose activity could not be mimicked by plasma proteinase inhibitors.

The experimental observations, as mentioned in this paragraph, clearly indicate that ALP is a highly potent inhibitor of neutrophil-proteinase-mediated matrix degradation and, when present in the parenchymal interstitium of the lung, may be of importance in regulating extracellular matrix destruction as occurs in emphysema.

Concluding Remarks

Circumstantial evidence has been obtained that pulmonary emphysema is the result of an unrestrained activity of elastase, mainly released by neutrophils. The severity of the disease may vary considerably between individuals suggesting that additional endogenous and exogenous factors, besides cigarette smoking and α_1PI deficiency, play a role in the pathogenesis of the disease. A factor which may limit the destructive effects of neutrophil elastase in the lung is the locally produced elastase-inhibitor, antileucoprotease. In addition to its ability to form rapidly a stable complex with free neutrophil elastase, ALP, as compared with α_1PI, proved to be a highly potent inhibitor of elastin-bound elastase and of matrix degradation mediated by adherent neutrophils. Evidence that ALP may play a role in protecting the lung parenchyma *in vivo* has been obtained by the observation that ALP is present along elastin fibres in the alveolar interstitium. Moreover, as observed by Willems et al.,[41] smoking related inflammation in peripheral airways and alveolar destruction was found to correlate with the local number of ALP-producing cells in the bronchiolar epithelium, suggesting involvement of the inhibitor in these processes.

Attention is directed nowadays to developing a preventive therapy of emphysema in α_1PI-deficient individuals.[42] The most obvious approach to augmenting the elastase-inhibitor screen in the lung of these individuals is the administration of a suitable inhibitor. Additionally, in the future this approach may also find application in the management of emphysematous patients with genetically sufficient α_1PI synthesis. The observation that ALP is capable of protecting very efficiently the extracellular matrix from elastase-mediated degradation suggests that this inhibitor has potential therapeutic value.

Antileucoprotease has been successfully produced by recombinant technology.[20,43,44] The recombinant protein, with an identical primary structure to native ALP, can be folded to a fully active inhibitor being composed of two separate domains each interconnected by four disulphide bridges. Aerosolization of recombinant ALP may prove to be very effective to augment the elastase inhibitor protective screen in the lung.

References

1. Janoff A.: Elastases and emphysema. Current assessment of the protease antiprotease hypothesis. Am. Rev. Respir. Dis. 1985; 132: 417-433
2. Weissler J.C.: Southwestern intestinal medicine conference: Pulmonary emphysema: Current concepts of pathogenesis. Am. J. Med. Sci. 1987; 293: 125-138
3. Snider G.L.: Chronic obstructive pulmonary disease: risk factors, pathophysiology and pathogenesis. Ann. Rev. Med. 1989; 40: 411-429
4. Janoff A., Pryor W.A., Bengali Z.H.: Effects of tobacco smoke components on cellular and biochemical processes in the lung. NHLBI Workshop Summary. Am. Rev. Respir. Dis. 1987; 136: 1058-1064
5. Sobonya R.E., Burrows B.: The epidemiology of emphysema. Clin. Chest Med. 1983; 4: 351-358
6. Mason R.J., Buist A.S., Fisher E.B., Merchant J.A., Samet J.M., Welsh C.H.: Cigarette smoking and health. Am. Rev. Respir. Dis. 1985; 132: 1133-1136
7. Janus E.D., Philips N.T., Carrell R.W.: Smoking, lung function, and alpha1-antitrypsin deficiency. Lancet 1985; i: 152-154
8. Schiessler H., Hochstrasser K., Ohlsson K.: Acid-stable inhibitors of granulocyte neutral protease in human mucous secretions: biochemistry and possible biological function. In: Havemann K., Janoff A. (Eds.): *Neutral proteases of human polymorphonuclear leukocytes*. Baltimore - Munich: Urban and Schwarzenberg, 1978; 195-207
9. Thompson R.C., Ohlsson K.: Isolation, properties, and complete amino acid sequence of human

120

secretory leucocyte protease inhibitor, a potent inhibitor of leucocyte elastase. Proc. Natl. Acad. Sci. USA 1986; 83: 6692-6696

10. Hochstrasser K., Reichert R., Schwarz S., Werle E.: Isolierung und Charakterisierung eines Proteaseninhibitors aus menshchlichem Bronchialsekret. Hoppe-Seyler's Z. Physiol. Chem. 1972; 353: 221-226

11. Ohlsson K., Tegner H., Akesson U.: Isolation and partial characterization of a low molecular weight acid stable protease inhibitor from human bronchial secretion. Hoppe -Seyler's Z.Physiol. Chem. 1977; 358: 583-589

12. Smith C.E., Johnson D.A.: Human bronchial leucocyte proteinase inhibitor. Rapid isolation and kinetic analysis with human leucocyte proteinases. Biochem. J. 1985; 225: 463-472

13. Fink E., Nettelbeck R., Fritz H.: Inhibition of mast cell chymase by Eglin c and antileukoprotease (HUSI-I). Indications for potential biological functions of these inhibitors. Biol. Chem Hoppe-Seyler 1986; 367: 567-571

14. Boudier C., Bieth J.G.: Mucus proteinase inhibitor: a fast acting inhibitor of leucocyte elastase. Biochem. Biophys. Acta 1989; 995: 36-41

15. Seemuller U., Arnhold M., Fritz H., Wiedenmann K., Machleidt W., Heinzel R., Appelhans H., Gassen H-G., Lottspeich F.: The acid-stable proteinase inhibitor of human mucous secretions (HUSI-I, antileucoprotease). FEBS - Letter 1986; 199: 43-48

16. Grütter M.G., Fendrich G., Huber R., Bode W.: The 2.5 A x-ray crystal structure of the acid-stable proteinase inhibitor from human mucous secretions analyzed in its complex with bovine alpha-chymotrypsin. EMBO J. 1988; 7: 345-351

17. Stetler G., Brewer M.T., Thompson R.C.: Isolation and sequence of a human gene enconding a potent inhibitor of leukocyte proteases. Nucleic. Acids Res. 1986; 14: 7883-7896

18. Kramps J.A., Twisk van Ch., Klasen E.C., Dijkman J.H.: Interactions among stimulated human polymorphonuclear leucocytes, released elastase and bronchial antileukoprotease. Clin. Sci. 1988; 75: 53-62

19. Kramps J.A., Twisk van Ch., Appelhans H., Meckelein B., Nikiforov T., Dijkman J.H.: Proteinase inhibitory activities of antileucoprotease are presented by its second COOH-terminal domain. Biochim. Biophys. Acta 1990; 1038:178-185

20. Eisenberg S.P., Hale K.K., Heidal P., Thompson R.C.: Location of the protease-inhibitory region of secretory leucocyte protease inhibitor. J. Biol. Chem. 1990; 265: 7976-7981

21. Kramps J.A., Franken C., Dijkman J.H.: Elisa for quantitative measurement of low-molecular-weight bronchial protease inhibitor in human sputum. Am. Rev. Respir. Dis. 1984; 129: 959-963

22. Franken C., Meijer C.J.L.M., Dijkman J.H.: Tissue distribution of antileukoprotease and lysozyme in humans. J. Histochem. Cytochem. 1989; 37: 493-498

23. Tegner H.: Quantitation of human granulocyte protease inhibitors in nonpurulent bronchial lavage fluids. Acta Otolaryngol. 1978; 85: 282-289

24. Kramps J.A., Franken C., Dijkman J.H.: Quantity of antileukoprotease relative to alpha1-proteinase inhibitor in peripheral airspaces of the human lung. Clin. Sci. 1988; 75: 351-353

25. Kramps J.A., Franken C., Meijer C.J.L.M., Dijkman J.H.: Localization of low molecular weight protease inhibitor in serous secretory cells of the respiratory tract. J. Histochem. Cytochem. 1981; 29: 712-719

26. De Water R., Willems L.N.A., Muijen van G.N.P., Franken C., Fransen J.A.M., Dijkman J.H., Kramps J.A.: Ultrastructural localization of bronchial antileukoprotease in central and peripheral human airways by a gold-labeling technique using monoclonal antibodies. Am. Rev. Respir. Dis. 1986; 133: 882-890

27. Willems L.N.A., Kramps J.A., de Water R., Stijnen Th., Fleuren G.J., Franken C., Dijkman J.H.: Evaluation of antileukoprotease in surgical lung specimens. Eur. J. Respir. Dis. 1986; 69: 242-247

28. Boudier C., Pelletier A., Gast A., Tournier J-M., Pauli G., Bieth J.G.: The elastase inhibitory capacity and the alpha1-proteinase inhibitor and bronchial inhibitor content of bronchoalveolar lavage fluids from healthy subjects. Biol. Chem. Hoppe-Seyler 1987; 368: 981-990

29. Willems L.N.A., Otto-Verbene C.J.M., Kramps J.A., Have ten-Opbroek A.A.W., Dijkman J.H.: Detection of antileukoprotease in connective tissue of the lung. Histochemistry 1986; 86: 165-168

30. Kramps J.A., Boekhorst te A.H.T., Fransen J.A.M., Ginsel L.A., Dijkman J.H.: Antileukoprotease is associated with elastin fibres in the extracellular matrix of the human lung. An immunoelectron microscopic study. Am. Rev. Respir. Dis. 1989; 140: 471-476

31. Bruch M., Bieth J.G.: Influence of elastin on the inhibition of leucocyte elastase by alpha1-proteinase inhibitor and bronchial inhibitor. Biochem. J. 1986; 238: 269-273

32. Morrison H.M., Welgus H.G., Stockley R.A., Burnett D., Campbell E.J.: Inhibition of human leukocyte elastase bound to elastin. Relative ineffectiveness and two mechanisms of inhibitory activity. Am. J. Respir. Cell Mol. Biol. 1990; 2: 263-269

33. Gauthier F., Fryksmark U., Ohlsson K., Bieth J.G.: Kinetics of the inhibition of leukocyte elastase by the bronchial inhibitor. Biochim. Biophys. Acta 1982; 700: 178-183

34. Hunninghake G.W., Crystal R.G.: Cigarette smoking and lung destruction. Accumulation of neutrophils in the lungs of cigarette smokers. Am. Rev. Respir. Dis. 1983; 128: 833-838

35. Senior R.M., Campbell E.J.: Neutral proteinases from human inflammatory cells. A critical review of their role in extracellular matrix degradation. Clin. Lab. Med. 1983; 3: 645-666

36. Weiss S.J.: Tissue destruction by neutrophils. N. Engl. J. Med. 1989; 320: 365-376

37. Hanlon W.A., Stolk J., Davies P., Bonney R.J.: Tumor necrosis factor alpha (rTNFalpha) facilitates the attachment of polymorphonuclear leucocytes to fibrinogen. J. Leuk. Biol. 1989; 46: 297 (abstract)

38. Mumford R.A., Andersen O.F., Boger J., Bondys S., Bonney R.J., Boulton D., Davies P., Doherty J., Fletcher D., Hand K., Mao J., Williams H., Dahlgren M.E.: Development of a direct, sensitive and specific RIA for a primary human leucocyte elastase (HLE) generated fibrinogen cleavage product, A−α (1-21). Am. Rev. Respir. Dis. 1989; 139: A574

39. Stolk J., Hanlon W.A., Davies P., Mumford R., Dahlgren M.E., Knight W.B., Kramps J.A., Bonney R.J.: The effect of antileucoprotease (ALP) and alpha1-proteinase inhibitor (alpha1PI) on the PMN-mediated degradation of fibrinogen. Am. Rev. Respir. Dis. 1989; 139: A201

40. Rice W.G., Weiss S.J.: Regulation of proteolysis at neutrophil-substrate interface by secretory leucoprotease inhibitor. Science 1990; 249: 178-181

41. Willems L.N.A., Kramps J.A., Stijnen Th., Sterk P.J., Weening J.J., Dijkman J.H.: Antileucoprotease-containing bronchiolar cells. Relationship with morphologic disease of small airways and parenchyma. Am. Rev. Respir. Dis. 1989; 139: 1244-1250

42. Gadek J.E., (Ed.): Alpha1-antitrypsin deficiency usage of alpha1-proteinase inhibitor concentrate in replacement therapy. Am. J. Med. 1988; 84: (6A), 1-90

43. Heinzel R., Appelhans H., Gassen G., Seemuller U., Machleidt W., Fritz H., Steffens G.: Molecular cloning and expression of cDNA for human antileucoprotease from cervix uterus. Eur. J. Biochem. 1986; 160: 61-67

44. Stetler G.L., Forsyth C., Gleason T., Wilson J., Thompson R.C.: Secretion of active, full- and half-length human secretory leukocyte protease inhibitor by saccharomyces cerevisae. Biotechnology 1989; 7: 55-60

10. Synthetic Mechanism-Based and Transition-State Inhibitors for Human Neutrophil Elastase

J. C. POWERS,[1] C.-M. KAM,[1] H. HORI,[1] J. OLEKSYSZYN,[1] E. F. MEYER JR.[2]

1. School of Chemistry, Georgia Institute of Technology, Atlanta, Georgia, USA
2. Department of Biochemistry, Texas A and M University, College Station, Texas, USA

Role of Elastase in Disease

Chronic Obstructive Pulmonary Disease (COPD), comprised of chronic bronchitis and pulmonary emphysema, is a serious health problem in the world. In the USA, COPD caused more than 70,000 deaths in 1986 and more than 10,000,000 Americans suffered from the disease.[1] The association between severe $\alpha_1$protease inhibitor (α_1PI) deficiency and emphysema has led to the hypothesis that an elastase-antielastase imbalance causes emphysema.[2,3] Human neutrophil elastase (HNE) is most likely the cause of emphysema in both smokers and non-smokers, although the evidence supporting the elastase-antielastase hypothesis is largely indirect in smokers with normal protective levels of α_1PI. Homogenates of leucocytes as well as highly purified preparations of HNE produce emphysema in experimental animals and only elastolytic enzymes will induce experimental emphysema. In addition, neutrophils are increased 4-5 fold in the lungs of smokers and it is postulated that α_1PI is inactivated by powerful oxidizing agents in the cigarette smoke and PMNs contributing to the elastase-antielastase imbalance. Other proteases including human proteinase 3 and macrophage elastase may also play a role in lung destruction in emphysema. Nevertheless, the majority of investigators in the field believe that a human neutrophil elastase-antielastase imbalance plays the major role in the pathogenesis of emphysema.

HNE is also involved in a variety of other pathological states including rheumatoid arthritis, adult respiratory distress syndrome, infantile respiratory distress syndrome, glomerulonephritis, cystic fibrosis, atherosclerosis, psoriasis, and bronchial secretory cell metaplasia. Bronchial secretory cell metaplasia has recently been shown to be induced by HNE[4] and is prevented by recombinant human secretory leucocyte protease inhibitor[5], a specific protein inhibitor of HNE. This raises the possibility that a protease-antiprotease imbalance results in chronic bronchial injury and suggests the lesion might be prevented by elastase inhibitor therapy. Cystic fibrosis (CF), the commonest potentially lethal genetic disease in the USA, is another disorder in which HNE has been implicated as a potential source of lung damage.

Large numbers of neutrophils occur in the airways and in the sputum. Neutrophil elastolytic activity is present and can destroy immunoglobulin molecules in the airways secretions. Desmosine, a marker of elastin destruction, occurs in increased amounts in the urine. Elastic fibre destruction occurs in the bronchi, and it is believed that the bronchiectasis characteristic of this disease may be caused by elastase-antielastase imbalance.[6] HNE has also been demonstrated in the bronchopulmonary secretions of patients with the adult respiratory distress syndrome[7,8] and it has been postulated that proteolytic enzymes may contribute to lung injury in that disease.

Clearly there is a need for effective elastase inhibitors for use in therapy to treat a variety of disease states involving lung injury. The great cost of emphysema in morbidity and mortality in the USA by itself warrants further research that would lead to the development of elastase inhibitors, which could be effectively administered either by aerosol or by mouth. The use of a synthetic elastase inhibitor which interfered with the process leading to emphysema in smokers would also provide direct evidence for the validity of the elastase-antielastase theory for pulmonary emphysema. In addition, elastase inhibitors are likely to be useful in the treatment of cystic fibrosis, adult respiratory distress syndrome, and bronchial secretory cell metaplasia.

The long term goal of our research is the design, synthesis, and testing of potent new low molecular weight synthetic elastase inhibitors suitable for use in the treatment of human pulmonary emphysema. Much of our current work is centred on mechanism-based (suicide or kamikaze) irreversible inhibitors for human neutrophil elastase.

Transition State Inhibitors

A variety of transition-state inhibitors for HN elastase have been developed.[9] Most so-called transition state inhibitors utilize the tetrahedral intermediate which is formed during peptide bond hydrolysis. Currently available transition state inhibitors include peptide aldehydes, peptide boronic acids,[10] and peptide trifluoromethyl ketones.[11] As shown in figure 1, these inhibitors form tetrahedral adducts which resemble the tetrahedral adduct formed during peptide bond hydrolysis.

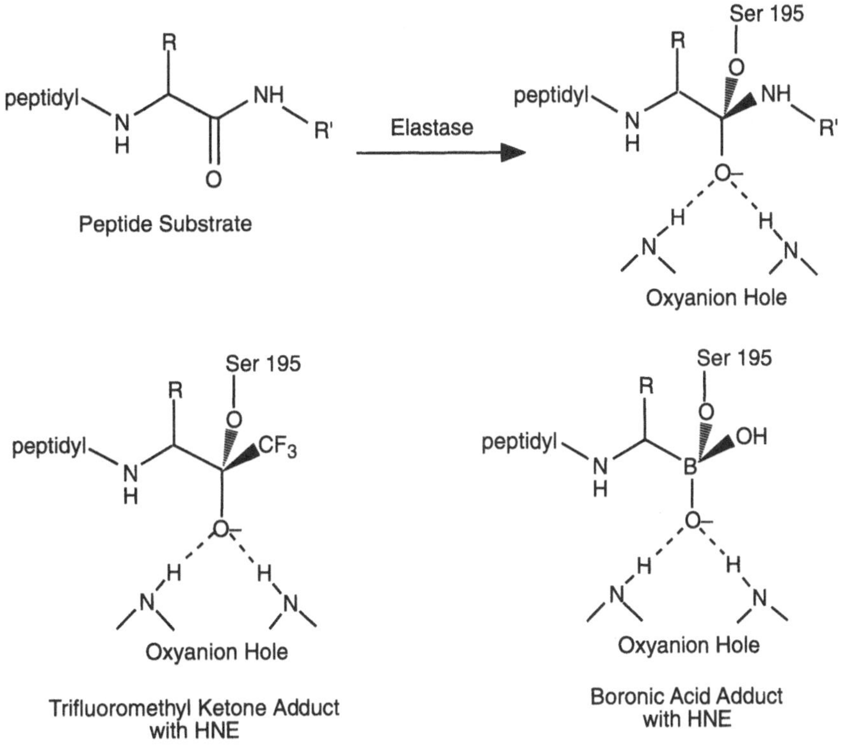

Fig. 1. Comparison of the tetrahedral adduct formed during peptide bond hydrolysis by HNE and the adducts formed upon inhibition of HNE with peptide trifluoromethyl ketone and peptide boronic acid inhibitors.

Peptide α-Ketoester Transition State Inhibitors

We have developed novel peptide α-ketoesters as another type of transition state inhibitor for elastases.[12] This class of inhibitors are based on the refined x-ray crystal structure of the complex formed between bovine trypsin and the ketoacid inhibitor 4-amidinophenylpyruvate (APPA), a potent trypsin inhibitor.[13] In the x-ray structure, the amidinophenyl group is located in the primary specificity pocket of trypsin and the active site serine has added to the α-carbonyl group in APPA to form a tetrahedral structure.

The oxyanion is stabilized by hydrogen bonding with groups in the oxyanion hole of trypsin. A unique feature of this structure is the hydrogen bonding observed between the carboxylate oxygen of APPA and the serine-195 oxygen and the NH of histidine-57.

We expected that the negative charge on the carboxylate would significantly contribute to the binding energy of α-ketoacid inhibitors.

We designed and synthesized a number of peptide derivatives of various α-

Table I. Inhibition of serine proteinases by Peptide α-Ketoacids and α-Ketoesters [a]

Inhibitor	HLE	PPE	Cat G
Bz-DL-Ala-COOEt	640	590	
Bz-DL-Ala-COOH	3100	3200	
Z-Ala-Ala-DL-Abu-COOEt	0.12	0.15	
Z-Ala-Ala-DL-Abu-COOBzl	0.09	0.08	
Z-Ala-Ala-Ala-DL-Ala-COOEt	0.3	0.14	
MeO-Suc-Val-Pro-DL-Phe-COOMe			1.1

[a] K_i values are reported in μM and were measured in a 0.1 M Hepes, pH 7.5, 0.5 M NaC1 buffer containing 9.0-9.8% Me$_2$SO at 25°C.

ketoacids and α-ketoesters derived from several different amino acids. Some of our kinetic results are shown in Table I. The α-ketoesters are abbreviated as peptidyl-AA-CO$_2$R and for example, Ala-CO$_2$Et = -NHCH (CH$_3$) CO-CO$_2$Et (Abu = 2-aminobutanoic acid). Surprisingly, the alpha-ketoesters are much better inhibitors than the corresponding α-ketoacids which indicates that additional interactions with the S$_1$' subsite and hydrogen bonding of the ester oxygen with histidine N-H can result in significant binding energy in the case of elastase.

The best HNE inhibitors, Z-Ala-Ala-DL-Abu-COOEt, the corresponding benzyl ester, and Z-Ala-Ala-Ala-DL-COOEt, had K_i values in the submicromolar range. The peptide ketoesters are quite specific for the target enzyme and for example the human neutrophil cathepsin G inhibitor MeO-Suc-Val-Pro-Phe-COOMe doesn't inhibit HNE or PPE (porcine pancreatic elastase). Importantly, MeO-Suc-Val-Pro-Phe-COOMe is one of the most potent reversible cathepsin G inhibitors known. Since our initial report[12], other investigators have also reported ketoester inhibitors for elastase.[14]

An x-ray crystal structure of an α-ketoester inhibitor bound to the active site of PPE has been completed by Ed Meyer and J. Vijayalakshmi (manuscript in preparation) at Texas A & M University and a schematic drawing of the interactions observed in the structure is shown in figure 2. The Ser-195 oxygen had added to the carbonyl group of the ketoester to form a tetrahedral intermediate which is stabilized by interactions with the oxyanion hole. This structure resembles the tetrahedral intermediate involved in peptide bond hydrolysis and proves that α-ketoesters are transition-state analogues. His-57 is hydrogen bonded to the carbonyl group of the ester functional group. The peptide backbone of a section of PPE's backbone hydrogen bonds to the inhibitor to form a β-sheet structure and the benzyl ester is directed toward the S' subsites. It appears likely that the potency of this class

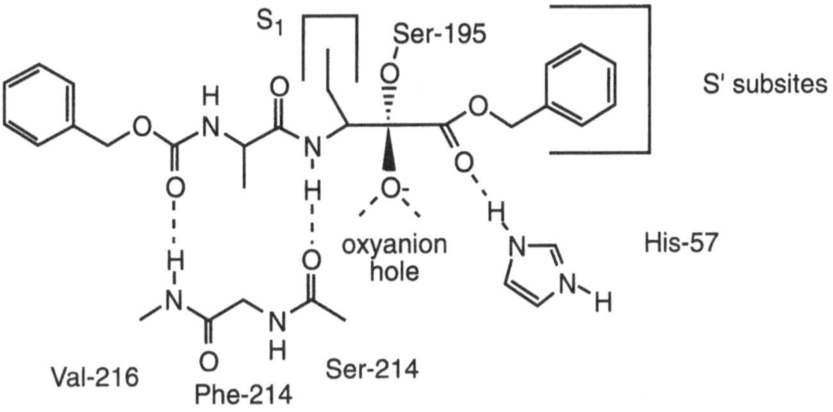

Fig. 2. Schematic drawing of the binding of the
α-ketoester inhibitor Z-Ala-Abu-CO$_2$Bzl to the active site of PPE. of PPE.

of inhibitor could be substantially improved if the structure was extended to allow for more interaction with the S' subsites.

Irreversible Inhibitors

A wide variety of irreversible inhibitors have been developed for serine proteases including HNE.[9] One group of inhibitors, often referred to as active site directed irreversible inhibitors, simply contain a substrate like structure which directs binding to the active site of HNE and a reactive group to react with an adjacent nucleophile. Examples of this type of inhibitor are peptide chloromethyl ketones (Fig. 3). Inhibitors such as MeO-Suc-Ala-Ala-Pro-Val-CH$_2$Cl are effective inhibitors of HNE and have been widely used in animal studies, but are alkylating agents and may be too toxic for the long term treatment of chronic diseases such as emphysema.

Another group of irreversible inhibitors are peptidyl derivatives of α-aminoalkylphosphonate diphenyl esters which we refer to as transition state irreversible inhibitors. These irreversible inhibitors are not alkylating agents and are more stable and less reactive in plasma than chloromethyl ketones.

Transition State Irreversible Inhibitors: Peptide Phosphonates

We have discovered that peptidyl derivatives of α-aminoalkylphosphonate diphenyl ester are effective and specific inhibitors of HNE and other serine proteases at low concentrations.[17,18] These inhibitors have a number of advantages which should make them useful for *in vivo* investigations. The compounds are

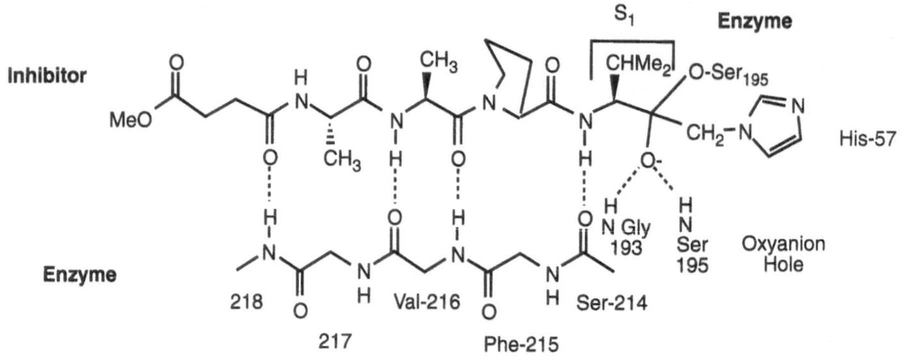

Fig. 3. Schematic drawing of the interaction of the peptide chloromethyl ketone inhibitor MeO-Suc-Ala-Ala-Pro-Val-CH$_2$Cl with the active site of HNE. The peptide chain of the inhibitor forms a β-sheet structure with the peptide backbone residues 214 to 218 of the enzyme. The imidazole ring of His 57 has displaced the chlorine from the chloromethyl ketone functional group and the Ser 195 has added to the inhibitor ketone carbonyl group to form an oxyanion which is stabilized in the oxyanion hole of the enzyme. The side chain of the valine residue occupies the S$_1$ pocket of HNE. The figure is based on two recent crystal structures of peptide chloromethyl ketone inhibitor complexes with HNE.[15,16]

chemically stable, relatively easy to synthesize, and do not react with acetylcholinesterase. A number of other phosphonate inhibitors of serine proteases, including phosphonyl fluorides or phosphonamides, have been synthesized previously[19,20] but they are unstable in solution and/or reactive toward acetylcholinesterase. Our phosphonates are stable for at least 3 days in human plasma and we believe that they have considerable utility as therapeutic agents due to their high stability and specificity.

The mechanism of inhibition of serine proteases by peptidyl derivatives of α-aminoalkylphosphonate diphenyl esters involves phosphonylation of the enzyme to form a stable phosphonyl enzyme derivative (Fig. 3). The inhibited enzyme derivative has a half-life for reactivation of 10 hrs in the case of chymotrypsin and >3 days in the case of HNE and PPE. The phosphonate diphenyl esters form very stable derivatives with serine proteases due to the resemblance between the inhibition product (phosphonyl derivative shown in figure 4) and the tetrahedral intermediate involved in peptide bond hydrolysis (Fig. 1).

The product phosphonate is tetrahedral and we believe that one of the phosphorus oxygens occupies the oxyanion hole. Thus we classify these inhibitors as transition state irreversible inhibitors.

Aminoalkyphosphonate diphenyl esters are very specific for the target serine protease and our data suggests that good interactions with the S$_1$ pocket of the target serine protease are necessary before nucleophilic substitution on phosphorus atom can occur to give a stable phosphonyl derivative. For example, Suc-Val-Pro-Phe^P (OPh)$_2$ reacts with chymotrypsin ($k_{obsd}/(I) = 44,000$ M^{-1}s^{-1}) and does not react with

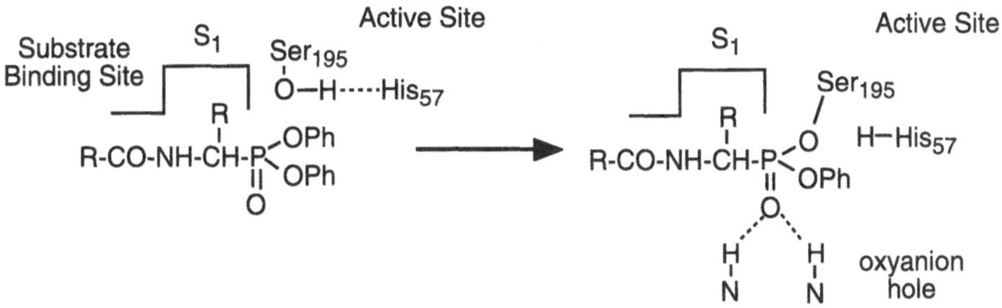

Active Site   Active Site

Substarte   S₁   Ser₁₉₅   S₁   Ser₁₉₅
Binding Site   O—H·····His₅₇   /   H—His₅₇
            R   _OPh   R   O
R-CO-NH-CH-P<   R-CO-NH-CH-P<
            ‖   OPh   ‖   OPh
            O   O
                                              ..O..
                                              H   H   oxyanion
                                              | | hole
                                              N   N

Fig. 4. Mechanism of inhibition of a serine protease by a peptidyl derivative of an α-aminoalkylphosphonate diphenyl ester.

elastase (Table II).[21] Boc-Val-Pro-Val^P (OPh)$_2$ reacts with HNE (27,000 $M^{-1}s^{-1}$) and PPE (11,000 $M^{-1}s^{-1}$) and does not react with chymotrypsin. The longer peptides with a C-terminal phosphonate related to phenylalanine are also inhibitors for other chymotrypsin-like enzymes such as rat mast cell protease II and human cathepsin G. The best phosphonate inhibitor of HNE is Boc-Val-Pro-Val^P (OPh)$_2$.

The inhibition reaction is stereospecific since ^{31}P NMR studies have shown that only one of the two stereo-isomers of Suc-Val-Pro-Phe^P (OPh)$_2$ reacts with chymotrypsin with a rate constant (146,000 $M^{-1}s^{-1}$) which is higher than the inhibition rate with the mixture. The ^{31}P NMR of chymotrypsin inhibited by this peptide shows one broad signal at 25.98 ppm corresponding to the Ser-195 phosphonate ester which is consistent with the product structure shown in figure 3.

Mechanism-Based Irreversible Inhibitors

Mechanism-based irreversible inhibitors are considered by many investigators to have much more potential therapeutic utility than active site directed irreversible inhibitors.[9] These inhibitors utilize the enzyme's catalytic mechanism in the inhibition reaction and often require enzyme catalysis to release a latent reactive group contained in the inhibitor structure before inactivation can occur. This later group of inhibitors are called suicide or kamikaze inhibitors. Mechanism-based inhibitors usually contain heterocyclic structures and have the advantage of being chemically relatively inactive until they reach the target enzyme. Many mechanism-based inhibitors utilize acyl enzyme formation in their inhibition mechanism and three examples are shown in figure 5. Benzoxazinone inhibitors[22,23,24] appear simply to form stable acyl enzyme structures with HNE. In contrast, many isocoumarin and β-lactam[25] inhibitors release or unmask other reactive groups upon acylation of HNE. Often these inhibitors will then form other covalent bonds with active site residues (usually His-57) and thus these inhibitors can be considered to be suicide or kamikaze inhibitors.

Table II. Inhibition of serine proteases by peptidyl derivatives of alpha-aminoalkylphosphonate diphenyl esters [a]

Inhibitor	ChyT	PPE	HNE
MeO-Suc-Ala-Ala-Pro-Nva^P (OPh)$_2$	50	4200	380
MeO-Suc-Ala-Ala-Ala-Val^P (OPh)$_2$	15	2800	1500
MeO-Suc-Ala-Ala-Ala-Phe^P (OPh)$_2$	2,000	NI	NI
MeO-Suc-Ala-Ala-Pro-Val^P (OPh)$_2$	21	7,100	7,100
MeO-Suc-Ala-Ala-Pro-Leu^P (OPh)$_2$	1,500	740	140
MeO-Suc-Ala-Ala-Pro-Phe^P (OPh)$_2$	11,000	NI	NI
MeO-Suc-Ala-Ala-Pro-Met^P (OPh)$_2$	570	44	53
MeO-Suc-Ala-Ala-Pro-Met(O)^P (OPh)$_2$	15	1.6	1.6
Boc-Val-Pro-Val^P (OPh)$_2$	NI	11,000	27,000
Z-Phe-Pro-Phe^P (OPh)$_2$	17,000	NI	NI
Suc-Val-Pro-Phe^P (OPh)$_2$	44,000	NI	NI

[a] The second inhibition rate constants $k_{obsd}/[I]$ are reported in $M^{-1}s^{-1}$ and were measured in a 0.1 M Hepes, pH 7.5, O.5 M NaCl buffer containing 9.0% Me_2SO at 25°C.

Isocoumarin Mechanism-Based Inhibitors

We have discovered that isocoumarins are mechanism-based heterocyclic serine protease inhibitors which are rich in possible masked reactive functional groups. Thus far, we have described isocoumarins which contain latent acid chloride (3,4-dichloroisocoumarin)[26] or quinone imine methide (7-amino-4-chloroisocoumarins) functional groups.

We have synthesized well over a hundred isocoumarin inhibitors and measured inhibition rate constants with a variety of serine proteases including HNE and PPE. The recent completion of four X-ray structures of complexes of PPE inhibited by isocoumarins has yielded important insights into the binding modes of these inhibitors and their mechanism of inactivation of PPE and HNE.

The inactivation mechanism of elastase by 3-alkoxy-7-amino-4-chloro-isocoumarins (1) is shown in figure 6.[27]

The active site Ser-195 attacks the isocoumarin carbonyl group and opens the isocoumarin ring to form an acyl enzyme (2). This reaction unmasks a latent quinone imine methide functional group (3) which is formed by the elimination of HCl from the acyl enzyme. This can react either with an enzyme nucleophile (His-57) to give an irreversibly inhibited enzyme structure (4) or with a solvent nucleophile to give a stable acyl enzyme (5). Partial reactivation by hydroxylamine with some inhibited derivatives suggests a partitioning between the two enzyme-inhibitor complexes (4 and 5) in solution with the nonreactivatable complex (4) containing

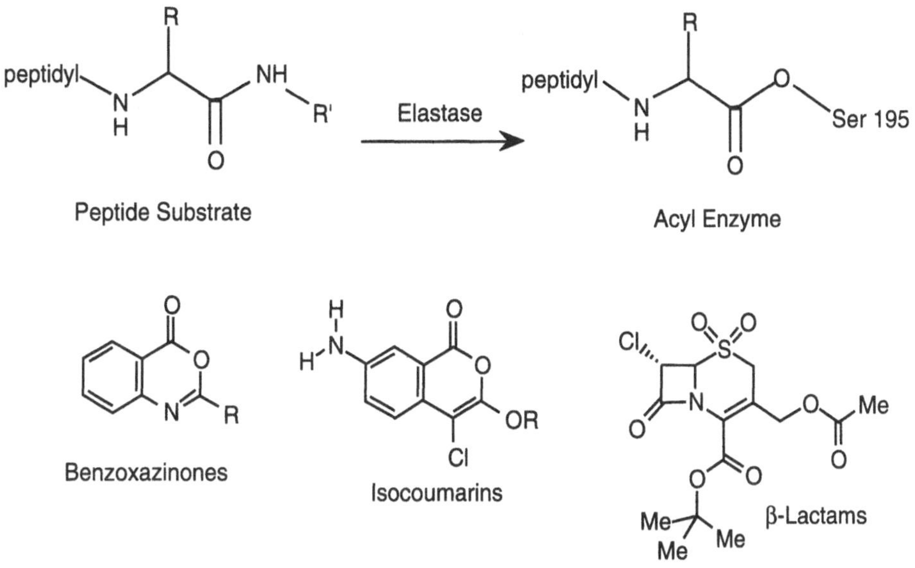

Fig. 5. The structure of the acyl enzyme formed during peptide bond hydrolysis by HNE and the structures of three classes of HNE inhibitors which utilize acyl enzyme formation in their inhibition mechanisms.

an alkylated histidine residue.

Both the 7-amino and 4-chloro groups are required for formation of a stable inactivated enzyme; isocoumarins which lack these features inhibit serine proteases, but deacylate fairly rapidly.

All the major products (2,4, and 5) in the reaction scheme have been observed crystallographically with different isocoumarins. The acyl enzyme structure 2 with the chlorine still present was found in the structure of PPE inhibited by 4-chloro-3-ethoxy-7-guanidinoisocoumarin.[28] The solvolysis product 5 (Nu = acetate) was observed in the structure of PPE inhibited by 7-amino-4-chloro-3-methoxyisocoumarin and PPE.[29]

The doubly covalent bound adduct 4 has been observed in the structure of trypsin inhibited by 4-chloro-3-ethoxy-7-guanidinoisocoumarin[30] and the structure of PPE inhibited by 7-amino-3-bromoethoxy-4-chloroisocoumarin. The observation of the doubly covalent bond histidine alkylation product 4 is the second time such a derivative has been observed with a serine protease. The first doubly covalent bound adduct was observed in a crystal structure of PPE inhibited by a β-lactam inhibitor which was reported by a Merck group.[31]

The exact product formed in the isocoumarin inhibition reaction varies with both the enzyme and the isocoumarin inhibitor. Different inhibitors give variable ratios of hydroxylamine reactivatable acyl enzymes (2 or 5) to non-reactivatable alkyla-

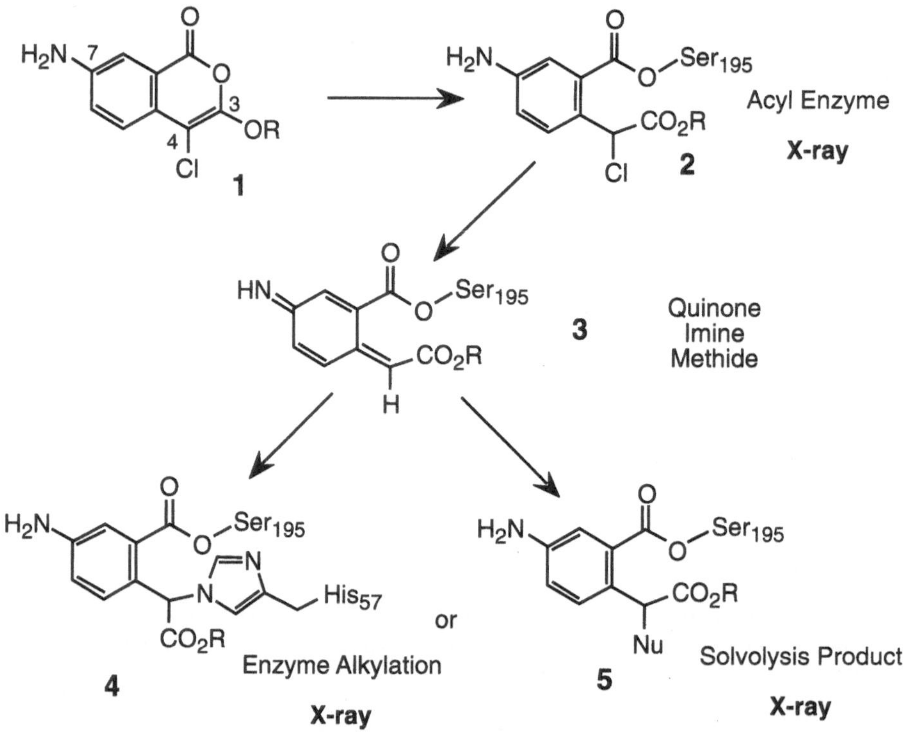

Fig. 6. Mechanism of inhibition of elastase by 3-alkoxy-7-amino-4-chloroisocoumarins. The label "X-ray" indicates that representative structures have been determined by X-ray crystallography for the indicated products or intermediates in the scheme.

tion products (**4**). The solvolysis product **5** is probably an artifact of that particular crystallographic experiment which was carried out in an acetate buffer at pH 5.0, a pH where the His-57 would be protonated and less likely to undergo alkylation. Kinetic studies with a variety of inhibitors indicate that histidine alkylation is usually the favoured product at neutral pH with both elastases and most of the isocoumarins studied.

Isocoumarin Binding Modes

The binding modes of the various isocoumarins complexed to PPE and trypsin are remarkably different even though all are tethered to Ser-195 via an ester bond. The carbonyl group of the ester bond linking Ser-195 to the inhibitor is in the oxyanion hole of the serine protease in the 7-amino-4-chloro-3-methoxyisocoumarin complex with PPE, while it is twisted out of the oxyanion hole in the 7-guanidino structure (Fig. 7). The twisting of the ester carbonyl group allows favourable hydrogen bonding between the 7-guanidino group and Thr-41 (**6**) and explains the stability of the complex toward deacylation. His-57 is either hydrogen bonded to the ester bond or is covalently linked with the inhibitor (**4**).

In all the PPE complexes, the 3-alkoxy group of the isocoumarin is either in the S_1O pocket or lying nearby (Fig. 7).

Once we realized that the 3-alkoxy group of the isocoumarin was interacting with the S_1O pocket of elastase, we synthesized a number of derivatives of 3-alkoxy-7-amino-4-chloroisocoumarin with 3-alkoxy groups of varying length. With HNE, we found the following second order inhibition constants k_{obsd}/[I]:MeO, 10,000; EtO, 9,400; PrO, 54,000; and bromoethoxy, 200,000 $M^{-1}s^{-1}$. Clearly HNE prefers the long alkoxy groups.

Utilization of Molecular Modelling for Inhibitor Design

Although it is not yet possible to predict the binding mode of a new isocoumarin inhibitor to elastase, molecular modelling with X-ray crystal structures is extremely useful for improving that inhibitor structure and for interpreting inhibition kinetic data obtained with related inhibitor structures.[32]

For example, molecular modelling of the 7-guanidinoisocoumarin PPE complex **6** (Fig. 7) suggested that the addition of a small alkyl group (*t*-butyl) to the guanidino group might increase affinity to the enzyme due to the presence of a small hydrophobic pocket near the terminal nitrogen of guanidino group and above Thr-41. Therefore, we synthesized a series of 7-alkyl-NH-CO-NH derivatives of 4-chloro-3-ethoxyisocoumarin.

These ureas were chosen due to the difficulty of synthesizing alkyl guanidino derivatives. Replacement of the 7-guanidino group by a urea functional group resulted in almost no loss of inhibitory potency (Table III). As predicted, the *t*-butyl-NH-CO-NH derivative was the most effective PPE inhibitor and had a second order inhibition rate k_{obsd}/[I] which was 3.5 fold higher than the parent inhibitor. Indeed, the compounds developed from the modelling are among the best irreversible inhibitors reported thus far for PPE.[28]

The structure determination of the complex of the 3-bromoethoxyisocoumarin (**7**, Fig. 6) with PPE has allowed us to interpret kinetic data obtained with PPE and the related inhibitor structures listed in Table IV (all the bromoethoxy derivatives are ca. 100 fold better with HNE, data not shown).

First it is clear that the bromopropoxy derivatives with a slightly long alkoxy group are over 100 fold poorer inhibitors of both PPE and HNE. All of the derivatives in which a hydrophobic group is placed on the 7-amino group are 3-10 fold better inhibitors than the parent with the exception of the phenyl urea derivative which is much poorer.

The 7-amino group of the isocoumarin points toward the S' subsites of PPE and HNE and we predict that the hydrophobic groups in the extended inhibitors are

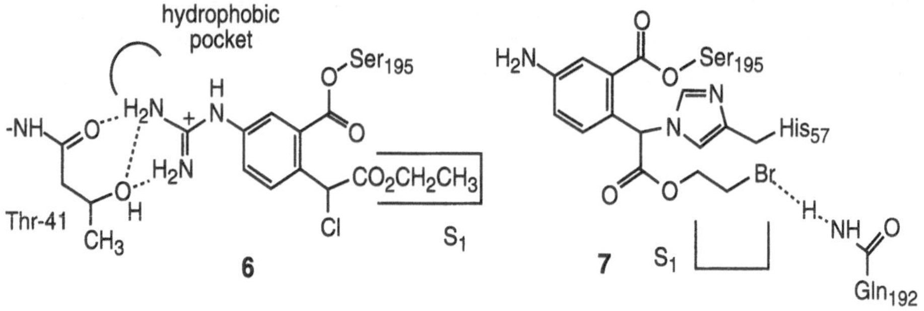

Fig. 7. Schematic drawing of the binding of 4-chloro-3-ethoxy-7-guanidinoisocoumarin 6 and 7-amino-3-(2-bromoethoxy)-4-chloroisocoumarin (**7**) to the active site of PPE.

interacting with S_2' subsite (Leu-143, Leu-151, Thr-41) in PPE. The phenyl urea derivative is probably so rigid, that the 7-substituent cannot twist and adopt a favourable conformation in the active site.

Human Neutrophil Elastase Inhibitors

Unfortunately, no crystal structure of an isocoumarin bound to HNE is available or likely to become available.[32] Several groups have tried extensively to crystallize complexes of HNE bound to small molecule inhibitors.

Up to the present, the smallest inhibitor which has yielded crystals with HNE is a tetrapeptide such as the chloromethyl ketone MeO-Suc-Ala-Ala-Pro-Val-CH$_2$Cl. However, the crystal structures obtained with PPE have been invaluable for modelling with HNE. The active site structures of PPE and HNE have many

Table III. Inhibition of porcine pancreatic elastase by 7-substituted derivatives of 4-chloro-3-ethoxyisocoumarin.[a]

7-Substituent	$k_{obsd}/[I]$ (M^{-1}s^{-1})
H$_2$N-C (=NH$_2^+$)-NH	2300
H$_2$N-CO-NH	2200
Me-HN-CO-NH	1400
Et-HN-CO-NH	1700
i-Pr-HN-CO-NH	4900
t-Bu-HN-CO-NH	8100
Ph-HN-CO-NH	4200

[a] The second inhibition rate constants were measured in a 0.1 M Hepes, pH 7.5, 0.5 M NaCl buffer containing 8.3% Me$_2$SO at 25°C.

Table IV. Inhibition of porcine pancreatic elastase by 7-substituted Derivatives of 3-(2-bromoethoxy)-4-chloroisocoumarin and 3-(3-bromopropoxy)-4-chloroisocoumarin.[a]

7-Substituent	$K_{obsd}/(I)$ $(M^{-1}s^{-1})$
3-(2-Bromoethoxy)-4-chloroisocoumarin	
7-NH$_2$	1,000
7-(Bu-NH-CO-NH)	6,600
7-(Ph-NH-CO-NH)	36
7-Ph-CH$_2$-NH-CO-NH)	3,010
7-(R-(C$_6$H$_5$) (CH$_3$)Ch-NH-CO-NH)	9,900
7-(Ph-Ch$_3$-CO-NH)	4,950
3-(3-Bromopropoxy)-4-chloroisocoumarin	
7-NH$_2$	10
7-(Ph-CH$_2$-NH-CO-NH)	13
7-(Ph-CH$_2$-CO-NH)	28

[a] The second inhibition rate constants were measured in a 0.1 M Hepes, pH 7.5, 0.5 M NaCl buffer containing 8.3% Me$_2$SO at 25°C.

features in common[32] and so we feel quite comfortable using the PPE crystal structure for modelling with HNE. For example, the complex of the 7-Tos-Phe derivative of 4-chloro-3-methoxy isocoumarin has been determined (J. Vijayalakshmi and E.F. Meyer, unpublished results) as has the 3-alkoxy group in the S$_1$ pocket and the Tos-Phe group interacting with the S' subsites of PPE (Fig. 8). We have overlaid this crystal structure into the HNE crystal structure and used it for molecular modelling.

The active site of HNE is much more hydrophobic than that of PPE.[32] Two significant changes are: Thr-41 in PPE is replaced in HNE by Phe-41 and Gln-192 is changed to Phe-192. Molecular modelling with HNE indicates that the Tos-Phe group is interacting with the S' subsites consisting of the side chains of Phe-41, Leu-35, and Leu-143. Isocoumarins with hydrophobic 7-substituents clearly should make favourable contacts with the S' subsites of HNE and as a result are extremely potent inhibitors. Representative inhibition rates with HNE are shown in Table V. The derivatives have a 3-propoxy group instead of the 3-methoxy group shown in the PPE crystal structure (Fig. 8) since we have demonstrated that HNE prefers an alkoxy group with two more methylene groups. Clearly, most of the rates are so fast that the rate constants should be considered to be lower limits.

Mechanism of Action of MR 889

MR 889 (Fig. 9), N-(S-(2-thiophenecarbonyl)-2-mercaptopropanoyl) homocysteine lactone, is a derivative of both homocysteine lactone and thiolactic acid.

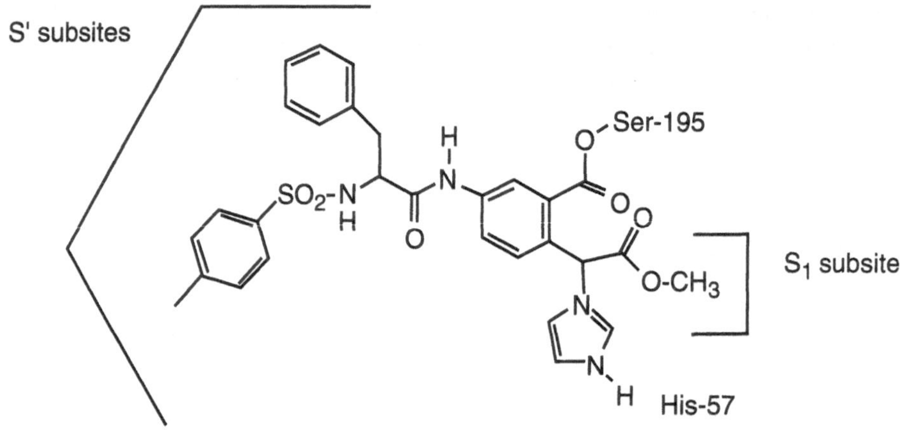

Fig. 8. Schematic drawing of the binding of the tosylphenylalanyl derivative of 7-amino-4-chloro-3-methoxyisocoumarin to the active site of PPE.

It has been reported to be an effective inhibitor of elastase and chymotrypsin.[33,34] Upon examination of the MR 889 structure, it appeared to us that the mechanism of inhibition would involve acylation of HNE by one of the two thioster functional groups in MR 889.

Table V. Inhibition of human neutrophil elastase by 7-substituted derivatives of 7-amino-4-chloro-3-propoxyisocoumarin[a]

7-substituent	$k_{obsd}/(I)$ (M^{-1}s^{-1})
Tos-Phe	33,400
CH$_3$CH$_2$CH$_2$CO	>124,000
(CH$_3$)$_2$CHCH$_2$CO	>220,000
PhCH$_2$CH$_2$CO	>250,000
3-O$_2$N-C$_6$H$_4$CO	>210,000
CH$_3$CH$_2$OCO	>181,000
CH$_3$CH$_2$HNCO	>276,000
PhNHCO	143,000
PhCH$_2$NHCS	>131,000
Ph-HNCS	>166,000

[a] The second inhibition rate constants were measured in a 0.1 M Hepes, pH 7.5, 0.5 M NaCl buffer containing 8.3% Me$_2$SO at 25°C.

Acylation of 2-thiophenecarbonyl thioester function group would yield a thiophenecarbonyl derivative of elastase (9) and the thiol (10). Alternatively, acylation of the enzyme by the homocysteine lactone functional group would yield the acyl enzyme 11. In both mechanisms, a new free thiol group would be generated (Fig. 9) and we decided to confirm thiol formation by titration with the thiol reagent 4,4'-dithiodipyridine (PDS). HNE was irreversibly inhibited by MR 889 in a 0.1 M Hepes, 0.5 M NaCl buffer at 25°C with $k_{obs}/(I)$ value of 270 $M^{-1}s^{-1}$. After 10 min incubation with an inhibitor concentration of 150 μM, the enzyme was 98% inhibited.

However, up to 40% of the activity was recovered during the assay of enzyme activity and the addition of 0.44 M hydroxylamine in buffer resulted in the recovery of 100% activity. These experiments are consistent with the formation of a rapidly deacylating acyl enzyme upon reaction of MR 889 with HNE. MR 889 is stable in the pH 7.5 Hepes buffer, shows no change in its UV spectrum after 1.5 hrs, and only a small change after 43 hrs. In addition, MR 889 (0.41 mM) does not react with the thiol reagent PDS (0.31 mM).

However, after treatment of MR 889 with 0.1 M NaOH, as expected two equivalents of thiol were released as measured by PDS titration. Reaction of MR 889 with 0.11 M hydroxylamine resulted in the release of 0.85 equivalent of thiol indicating that one of the two thioester functional groups in MR 889 is more reactive than the other. After reaction with HNE, titration with PDS yielded 0.87-1.0 equivalents of thiol which is consistent with formation of either of the two acyl enzymes shown in figure 9.

The MR 889 analog 8, in which one of the thioester linkages is replaced with a peptide bond, is not an inhibitor of HNE, PPE, chymotrypsin, or cathepsin G (4-13% inhibition after incubation for 15 min). In addition, the peptidyl homocysteine lactone derivative 12 at 0.21-0.38 mM, inhibited PPE and HNE only 12-18% after 30 min of incubation and did not inhibit chymotrypsin or cathepsin G. Since both 8 and 12 have a homocysteine lactone functional group and are not inhibitors of HNE, we propose that the most likely mechanism of inhibition of HNE by MR 889 involves formation of the thiophene derivative 9. This derivative would be expected to be a fairly unstable derivative and to deacylate rapidly upon standing.

Summary and Perspectives

A variety of elastase inhibitors has been shown to be effective in animal models of pulmonary emphysema, inflammation, and other related diseases.[35] These include peptide chloromethyl ketones,[36] peptide aldehydes, β-lactams, and the protein protease inhibitor, eglin. Human α_1 protease inhibitor has already been used in therapy with PiZ (α_1 protease inhibitor deficiency) patients.[37] Several additional

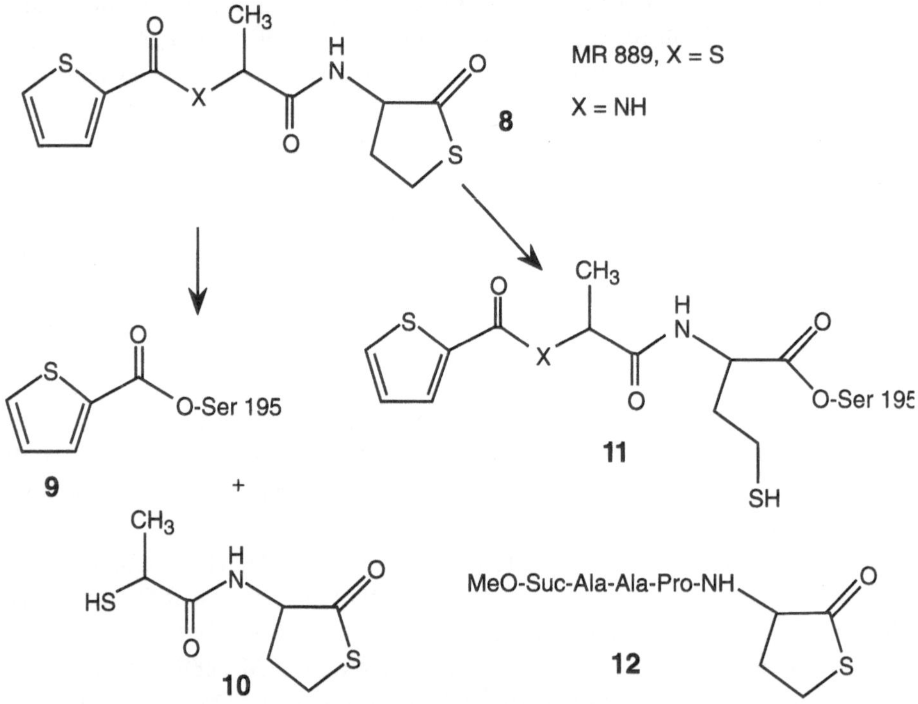

138

Fig. 9. Structure of MR 889 and related thiolactones. Possible reaction products formed upon inhibition of elastase by MR 889 are also shown.

inhibitors will be tested in the near future for treatment of disease in humans. At present it is most likely that the first practical therapeutic drugs will originate from either the β-lactam, or peptide fluoroketone classes of elastase inhibitors since current classes of heterocyclic inhibitors suffer from low plasma stability. The structural information obtained with elastase inhibitors complexes should be invaluable for future design work with all classes of elastase inhibitors and should improve the prospects for the treatment of chronic diseases such as pulmonary emphysema.

Acknowledgments and Dedication

This research was supported by a grant from the National Institutes of Health (HL 29307). The authors also wish to thank Dr. James Travis and his research group at the University of Georgia for the neutrophil enzymes used in this research and for many hours of valuable and stimulating discussions.
We dedicate this manuscript to the memory of Dr. Giorgio Staibano who is missed by his family and colleagues.

References

1. Higgins M.W., Thom T.: Incidence, prevalence and mortality: intra and intercounty differences. In: Hensley M.J., Saunders N.A. (Eds.). *Clinical epidemiology of chronic obstructive pulmonary Disease*. New York, Marcel Dekker Inc., 1989; 24-30

2. Janoff A.: Elastase and emphysema. Current assessment of the protease-antiprotease hypothesis. Am. Rev. Respir. Dis. 1985; 132: 417-411

3. Snider G.L., Lucey E.C., Stone P.J.: Animal models of emphysema. Am. Rev. Respir. Dis. 1986; 133: 149-169

4. Snider G.L., Lucey E.C., Christensen T.G., Stone P.J., Calore J.D., Catanese A., Franzblau C.: Emphysema and bronchial secretory cell metaplasia induced in hamsters by human neutrophil products. Am. Rev. Respir. Dis. 1984; 129: 155-160

5. Lucey E.C., Stone P.J., Ciccolella D.E., Breuer R., Christensen T.G., Thompson R.C., Snider G.L.: Recombinant human leukocyte-protease inhibitor ameliorates human neutrophil elastase induced emphysema and secretory cell metaplasia in the hamster. J. Lab. Clin. Med. 1990; 115: 224-232

6. Bruce M.C., Ronez L., Kilinger J.D.: Biochemical and pathologic evidence for proteolytic destruction of lung connective tissue in cystic fibrosis. Am. Rev. Respir. Dis. 1985; 132: 529-535

7. Lee C.T., Fein A.M., Lippmann M., Holtzman H., Kimbel P., Weinbaum G.: Elastolytic activity in pulmonary lavage fluid from patients with adult respiratory-distress syndrome. N. Engl. J. Med. 1981; 304: 192-196

8. Cochrance C.G., Spragg R.G., Revak S.D., Cohen A.B., McGuire W.W.: The presence of neutrophil elastase and evidence of oxidation activity in bronchoalveolar lavage fluid of patients with adult respiratory distress syndrome. Am. Rev. Respir. Dis. 1983; 127: S25-S27

9. Powers J.C., Harper J.W.: Inhibitors of serine proteases. In: Barrett A.J., Salvensen G.S. (Eds.). *Proteinase inhibitors*. Amsterdam/New York, Elsevier Science Publishers, 1986; 55-152

10. Kettner C.A., Shenvi A.B.: Inhibition of the serine proteases leukocyte elastase, pancreatic elastase, cathepsin G., and chymotrypsin by peptide boronic acids. J. Biol. Chem. 1984; 259: 15106-15114

11. Trainor D.A.: Synthetic inhibitors of human neutrophil elastase. Trends in Pharmacol. Sci. 1987; 87: 303-307

12. Hori H., Yasutake A., Minematsu Y., Powers J.C.: Inhibition of human leuckocyte elastase, porcine pancreatic elastase and caphepsin G by peptide ketones. In: Deber C.M., Hruby V.J., Kopple K.D. (Eds.). *Peptides: synthesis-structure-function. Proceedings of the ninth american peptide symposium*. Pierce Chem. Co., II, 1985; 819-822

13. Walter J., Bode W.: The x-ray crystal structure analysis of the refined complex formed by bovine trypsin and p-amidinophenylpyruvate at 1.4 A resolution. Hoppe-Seyler's Z. Physiol. Chem. 1983; 364: 949-959

14. Peet N.P., Brurkhart J.P., Angelastro M.R., Giroux E.L., Mehdi S., Bey P., Kolb M., Neises B., Schirlin D.: Synthesis of peptidyl fluoromethyl ketones and peptidyl alpha-keto esters as inhibitors of porcine pancreatic elastase, human neutrophil elastase, and rat and human neutrophil cathepsin G.J. Med. Chem. 1990; 33: 394-407

15. Bode W., Wei A-Z., Huber R., Meyer E., Travis J., Neumann S.: X-ray structure of the complex of human leukocyte elastase (PMN elastase) and the third domain of the turkey ovomucoid inhibitor. EMBO J. 1986; 5: 2453-2458

16. Navia M.A., McKeever B.M., Springer J.P., Lin T-Y., Williams H.R., Fluder E.M., Dorn C.D., Hoogsteen K.: Structure of human neutrophil elastase in complex with a peptide chloromethyl ketone inhibitor at 1.84-A resolution. Proc. Natl. Acad. Sci. U.S.A. 1989; 86: 7-11

17. Oleksyszyn J., Powers J.C.: Irreversible inhibition of serine proteases by peptidyl derivatives of alpha-aminoalkylphosphonate diphenyl esters. Biochem. Biophys. Res. Commun. 1989; 161: 143-149

18. Oleksyszyn J., Powers J.C.: Irreversible inhibition of serine proteases by peptide derivatives of alpha-aminoalkylphosphonate diphenyl esters. Biochemistry 1991; 30: 485-493

19. Bartlett P.A., Lamden L.A.: Inhibition of chymotrypsin by phosphonate and phosphonamidate analogs. Bioorg. Chem. 1986; 14: 356-377

20. Lamden L.A., Bartlett P.A.: Aminoalkylphosphonofluoridate derivatives: rapid and potentially selective inactivators of serine proteases. Biochem. Biophys. Res. Commun. 1983; 112: 1085-1090

21. The alpha-aminoalkylphosphonic acids are analogues of natural alpha-amino acids and are designated by the generally accepted three letter abbreviations for the amino acid followed by the superscript P. For example dephenyl alpha-(N-benzyloxycarbonylamino) ethylphosphonate which is related to alanine is abbreviated as Z-Ala^P (OPh)$_2$.

22. Teshima T., Griffin J.C., Powers J.C.: A new class of heterocyclic serine protease inhibitors. Inhibition of human leukocyte elastase, porcine pancreatic elastase, cathepsin G, and bovine chymotrypsin alpha with substituted benzoxazinones, quinazolines, and anthranilates. J. Biol. Chem. 1982; 257: 5085-5091

23. Krantz A., Spencer R.W., Tam T.F., Thomas E., Copp L.J.: Design of alternate substrate inhibitors of serine protease: synergistic use of alkyl substitution to impede enzyme-catalyzed deacylation. J. Med. Chem. 1987; 30: 589-591

24. Krantz A., Spencer R.W., Tam T.F., Liak T.J., Copp L.J., Thomas E.M., Rafferty S.P.: Design and synthesis of 4H-3,1-benzoxazin-4-ones as potent alternate substrate inhibitors of human leukocyte elastase. J. Med. Chem. 1990; 33: 464-479

25. Doherty J.B., Ashe B.M., Argenbright L.W., Barker P.L., Bonney R.J., Chandler G.O., Dahlgren M.E., Dorn C.P., Finke P.E., Firestone R.A., Fletcher D., Hagmann W.K., Mumford R., O'Grady L., Maycock A.L., Pisano J.M., Shah S.K., Thompson K.R., Zimmerman M.: Cephalosporin antibiotics can be modified to inhibit human leukocyte elastase. Nature 1986; 322: 192-194

26. Harper J.W., Hemmi K., Powers J.C.: Reaction of serine proteases with substituted isocoumarins: discovery of 3,4-dichloroisocoumarin, a new general mechanism based serine protease inhibitor. Biochemistry 1985; 24: 1831-1841

27. Harper J.W., Powers J.C.: Reaction of serine proteases with substituted 3-alkoxy-4-chloroisocoumarins and 3-alkoxy-7-amino-4-chloroisocoumarins: new reactive mechanism-based inhibitors. Biochemistry 1985; 24: 7200-7213

28. Powers J.C., Oleksyszyn J., Narasimham S.L., Kam C-M., Radhakrishnan R., Meyer E.F.: Reaction of porcine pancreatic elastase with 7-substituted 3-alkoxy-4-chloroisocoumarins: design of potent inhibitors using the crystal structure of the complex formed with 4-chloro-3-ethoxy-7-guanidinoisocoumarin. Biochemistry 1990; 29: 3108-3118

29. Meyer E.F., Presta L.G., Radhakrishnan R.: Stereospecific reaction of 3-methoxy-4-chloro-7-aminoisocoumarin with crystalline porcine pancreatic elastase. J. Am. Chem. Soc. 1985; 107: 4091-4093

30. Chow M.M., Meyer E.F., Bode W., Kam C-M., Radhakrishnan R., Vijayalashmi J., Powers J.C.: The 2.2 Å resolution x-ray crystal structure of the complex formed by reaction of the thrombin inhibitor 4-chloro-3-ethoxy-7-guanidinoisocoumarin with trypsin. J. Am. Chem. Soc. 1990; 112: 7783-7789

31. Navia M.A., Springer J.P., Lin T-Y., Williams H.R., Firestone R.A., Pisano J.M., Doherty J.B., Finke P.E., Hoogsteen K.: Crystallographic study of a β-lactam inhibitor complex with elastase at 1.84 Å resolution. Nature 1987; 327: 79-82

32. Bode W., Meyer E., Powers J.C.: Human leukocyte and porcine pancreatic elastase: x-ray crystal structures, mechanism, substrate specificity, and mechanism-based inhibitors. Biochemistry 1989; 28: 1951-1963

33. Baici A., Pelloso R., Horler D.: The kinetic mechanism of inhibition of human leukocyte elastase by MR 889, a new cyclic thiolic compound. Biochem. Pharmacol. 1990; 39: 919-924

34. Luisetti M., Piccioni P.D., Donnini M., Peona V., Pozzi E., Grassi C.: Studies of MR 889, a new synthetic proteinase inhibitor. Biochem. Biophys. Res. Commun. 1989; 165: 568-573

35. Powers J.C., Bengali Z.H.: Elastase inhibitors for treatment of emphysema. Am. Rev. Respir. Dis. 1986; 134: 1097-1100

36. Stone P.J., Lucey E.C., Calore J.D., Snider G.L., Franzblau C., Castillo M.J., Powers J.C.: The moderation of elastase-induced emphysema in the hamster by intratracheal pretreatment or posttreatment with succinyl alanyl prolyl valine chloromethyl ketone. Am. Rev. Respir. Dis. 1981; 124: 56-59

37. Gadek J.E., Klein H.G., Holland P.V., Crystal R.G.: Replacement therapy of alpha 1-antitrypsin deficiency. Reversal of protease-antiprotease imbalance with the alveolar structures of PiZ subjects. J. Clin. Invest. 1981; 68: 1158-1165

11. Development and Evaluation of Antiproteases as Drugs for Preventing Emphysema

G. L. SNIDER, P. J. STONE, E. C. LUCEY
The Boston Veterans Administration Medical Center; Pulmonary Center and Biochemistry Department, Boston University School of Medicine; Boston University and Tufts University Schools of Medicine, Boston Massachusetts, USA

Introduction

Emphysema is a process characterized by enlargement of the respiratory airspaces of the lungs with destruction of the walls between airspaces. In persons with adequate protective levels of α-1 protease inhibitor (API) the most important risk factors for the development of emphysema are the smoking history, age and the level of ventilatory function as measured by the forced expiratory volume for 1 sec (FEV$_1$).[1-4] About 15% of long term smokers are at risk of developing chronic obstructive pulmonary disease (COPD).[3]

The prevalence of smoking in the United States in persons age 20 and over has fallen from 52.1% in 1965, the year after the first report of the Surgeon General on Smoking and Health was issued to 32.7% in 1985[5] and 26.5% in 1986.[6]

Nevertheless, the last figure represents about 50 million adult Americans who are still smoking, of whom about 7 million are at risk of developing COPD. There are currently about 13 million sufferers from COPD in the US and about 70,000 deaths from the disease per year.

This public health problem is not unique to the United States. In 1985, the apparent annual consumption of cigarettes per adult (persons aged 15 years and over) was similar to the United States (>2500 *per capita*) in 18 nations: Australia, Canada, Cuba, Cyprus, Greece, Hungary, Iceland, Ireland, Japan, Kuwait, Lebanon, Libya, New Zealand, Poland, Republic of Korea, Spain, Switzerland and Yugoslavia.[7]

Smoking cessation is the obvious solution to prevention of emphysema. However, since tobacco smoking is highly addictive, about 75% of persons are still smoking one year after participation in intensive programmes of smoking cessation.[8] This paper will address the question of how antiproteases might be developed as drugs to prevent emphysema in people who cannot stop smoking. We shall emphasize, the type of *in vitro* and *in vivo* tests that might be used in the preclinical phases of drug development and the steps that might be taken in order to establish efficacy by clinical testing.

The Elastase-Antielastase Hypothesis

The elastase-antielastase hypothesis of pathogenesis of emphysema[9-11] holds that the lung destruction of emphysema occurs in smokers because their antielastase defences cannot adequately protect the elastic fibres of their lungs from damage by their own elastase. There is direct evidence for this hypothesis in persons with homozygous API deficiency. The evidence is indirect but compelling in smokers with adequate protective levels of API. Lung neutrophils are believed to be the most likely source of elastase. Their percentage in bronchoalveolar lavage fluid of smokers is unchanged, but their absolute number is increased because of a 4-5 fold increase in total leucocyte number.

Macrophages have also been implicated; they can ingest and later release neutrophil elastase. Macrophages also make a metalloelastase and are capable of digesting elastic fibres with which they are in contact.[12,13]

The major antiprotease in the lungs is API. Oxidation of this protein by the oxidants in cigarette smoke or resulting from the myeloperoxidase system of the neutrophil, results in its inactivation.

There is controversy regarding the in vivo evidence for oxidation of API in human smokers. Bronchial secretory leucocyte protease inhibitor (SLPI) normally present in airways secretions and the alveolar walls,[14] and α-2 macroglobulin, which can enter the lung when capillary permeability is increased, also play a role in the antiprotease defences of the lungs.

Preventive Strategies

The elastase-antielastase hypothesis suggests a number of different strategies that might be pursued in developing drugs for emphysema prevention. These can be categorized into antioxidants, designed to prevent oxidation of antiproteases, attempts to decrease the elastase burden of the lungs and attempts to supplement the antielastase of the lungs.

Antioxidants

Little work has been done with antioxidants,[15-17] perhaps because the evidence for the importance of this mechanism is controversial.[18-22]

Altering the elastase burden

Cohen has developed a strategy to decrease the elastase load of the lungs.[23] He has pointed out that there are many drugs currently on the market that affect neutrophil function without increasing the risk of infection; these might be used to prevent neutrophils from entering the lungs or discharging their enzymes. However he was unable to show in a controlled trial in smokers that colchicine, which prevents degranulation of neutrophils by preventing microtubule assembly, lowers the elastase load in the lungs or affects markers of elastin degradation.[24]

Elastase inhibitors

Another approach is to supplement the naturally occurring elastase inhibitors in the lungs by treatment with a naturally occurring or synthetic elastase inhibitor. These compounds have held the most interest for investigators in both academe and industry and we will review our own experience with some of these agents.

In vitro *studies*

In vitro studies consist of studies of solubility in saline, and the antielastase potency of the inhibitor in a system using HNE and tritiated elastin as substrate.[25] Several concentrations of the inhibitor are used and the molar ratio of inhibitor to HNE for 50% inhibition of HNE (HNE-MRI$_{50}$) is calculated. The half-time of dissociation of the HNE-inhibitor complex (HNE-Dt$_{1/2}$) is determined in a separate experiment.

In vivo *studies*

In vivo experiments test the ability of the preparation to ameliorate emphysema induced by elastase in hamsters.[26,27] Hamsters are given an intratracheal dose of the inhibitor to be tested, followed 1 to 16 hours later by an intratracheal dose of human neutrophil elastase (HNE).

Thirty to 56 days later, lung volumes and compliance are measured under anaesthesia, the lungs are prepared for histologic examination and the mean linear intercept (MLI) a measure of airspace size, is determined. In some experiments with rapidly clearing, reversible inhibitors, when there is a possibility of worsening of emphysema, groups of animals are treated with a mixture of HNE and an excess of inhibitor.

The half-time of lung clearance (LCt$_{1/2}$) of the inhibitors is also measured. Hamsters are instilled with a dose of the inhibitor. Groups of animals are killed at various times after the instillation, bronchoalveolar lavage is performed, and the amount of inhibitor remaining is determined from the HNE-inhibiting potency of the lavage fluid; a correction is made for the inhibiting potency of control lavage fluid. From these data we calculated the *in vivo* HNE-MR$_{50}$, the estimated molar ratio of lavageable inhibitor to instilled HNE for 50% amelioration of emphysema.[28]

Functional classification of elastase inhibitors

Elastase inhibitors include small molecular weight synthetic compounds and larger molecules such as API, eglin-c and SLPI, of natural or recombinant DNA origin.

We have proposed a functional classification of inhibitors based on the relations between their *in vitro* and *in vivo* properties (Table I).

Table I. *In vitro* and *in vivo* properties of 11 elastase inhibitors

Inhibitor	Mol. wt.	HNE-MR$_{50}$		Dt$_{1/2}$	LCt$_{1/2}$
		In vitro	*In vivo*		
Cat 1 SULF[58]	214	*	*	nd	nd
SACCH[59]	277	*	*	5,7d	nd
COUM[60]	180	24	**	2 h	nd
ABU[61]	435	29	***	nd	nd
Cat 2 CMK[62]	488	3.0	8.9	irrevers.	4.5 m
API[28]	53.000	0.78	4.8	irrevers.	4 h
Cat 3 Eglin-c[26,27]	8.100	0.5	1.25	5,5 h	35 m
SLPI[63]	11.700	0.55	2.3	1,8 h	2 h
PPCI[34]	22.000	0.38	0.37	14 min	7.6 h
Cat 4 BOROVAL[30,31]	553	1.5	****	30 min	15 m
PCI[33]	591	4.4	****	nd	3-6 m

Abbreviations: *In vitro*, HNE-MR$_{50}$: molar ratio of inhibitor to human neutrophil elastase (HNE) for 50% inhibition; *In vivo* HNE-MR$_{50}$: estimated molar ratio of inhibitor to HNE for 50% amelioration of emphysema; Dt$_{1/2}$: half-time for dissociation of HNE-inhibitor complex; LCt$_{1/2}$: half-time of clearance of inhibitor from the lavageable compartment of the lungs; SULF: 2-(CF$_3$CONH)-C$_6$H$_4$-SO$_2$F; SACCH: furoyl saccharin; COUM: 3-chloroisocoumarin; ABU: benzyloxicarbonyl-Ala-Ala-D,L-aminobutanoic-CO$_2$Et; CMK: succinyl-Ala-Ala-Pro-Val-chloromethyl ketone; API: alpha-1-protease inhibitor; SLPI: secretory leucocyte protease inhibitor; PPCI: peptidyl carbamate inhibitor (methoxysuccinyl-L-alanyl-L-alanyl-L-prolyl-CH$_2$N(i-Pr)CO$_2$-p-nitrophenol) bound to a water soluble polymer (N(2-hydroxyethyl)-D,L-aspartamide); BOROVAL: methoxy succinyl-L-alanyl-L-alanyl-L-prolyl-ambo-boro-val pinacol; PCI: peptidyl carbamate inhibitor (methoxysuccinyl-L-alanyl-L-alanyl-L-prolyl-CH$_2$N(i-Pr)CO$_2$-p-nitrophenol);nd: not done; Cat: category; category 1, sparingly soluble inhibitors; category 2, irreversible inhibitors; category 3, tight-binding, slowly clearing inhibitors; category 4, reversible, rapidly clearing inhibitors.

*1mg of SULF or SACCH, made up in a slurry with phosphate-buffered saline, completely inhibited 10μg of HNE; 1mg of each agent was administered as a slurry 1h before HNE and did not ameliorate emphysema.

**, Due to solubility limitations, 0.2mg of COUM was given 1h before HNE (107:1 molar ratio); there was no amelioration of emphysema.

***, ABU 1 mg, dissolved in 0.5 ml of 6.6% DMSO in saline, given 1h before HNE did not ameliorate emphysema.

****, Emphysema worsens

Category 1. Sparingly soluble agents

Sparingly water-soluble agents (approximately 1 mM or less) are generally ineffective, possibly because, relative to inhibitory potency, adequate amounts of the inhibitor are not in solution in the lungs. It may also be that lipophilic compounds clear rapidly from the lungs.[29] Attempts to increase bioavailability and *in vivo* effectiveness by dissolving the agents in 5-10% dimethyl sulfoxide (DMSO) solutions have generally been unsuccessful.

Category 2. Irreversible inhibitors

This category includes irreversible inhibitors such as oligopeptide chloromethyl ketones and native[28] and recombinant API, which are effective for many hours in ameliorating emphysema (Fig. 1). They do so in a dose dependent fashion.

Category 3. Tight-binding slowly clearing inhibitors

This category includes tight-binding, high molecular weight inhibitors, such as eglin-c[26,27] and SLPI that clear slowly from the lungs. These agents also prevent

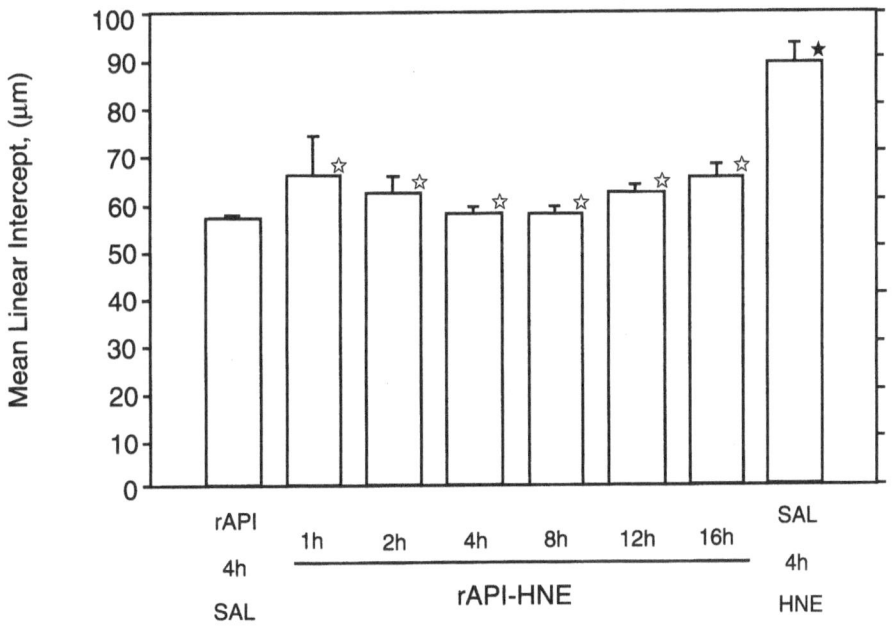

Fig. 1. Duration of effect of recombinant α_1-protease inhibitor (rAPI) in protecting against emphysema induction by human neutrophil elastase (HNE). Groups of hamsters received 4.3 mg of rAPI intratracheally and 1-16 hours later were given 300 µg of HNE. At 56 days the mean linear intercept of the lungs, a measure of airspace size, was determined. The positive control (saline followed in 4 hours by HNE) was the only group that was significantly different (★, $p < 0.05$) from the negative control group (rAPI followed in 4 hours by saline). All rAPI groups out to 16 hours were significantly different from the positive control group (☆, $p < 0.05$) indicating the long-persisting effectiveness of this agent in protecting against emphysema induction by HNE.

HNE-induced emphysema in a dose dependent fashion and do so for many hours.

Category 4. Reversible rapidly clearing inhibitors

Reversible, rapidly clearing inhibitors have the potential for worsening emphysema. The basis for this statement is our study, recently reported,[30] of a reversible small molecular weight inhibitor, methoxy succinyl-L-alanyl-L-alanyl-L-prolyl-ambo-boro-val pinacol (Boroval).

Boroval is a highly effective *in vitro* inhibitor of HNE[30] and PPE[31] and is effective in preventing the induction of emphysema in the hamster by pancreatic elastase.[31,32] However, it is not effective against the induction of emphysema by HNE.[30] Treatment of hamsters with as much as 3000 µg of Boroval 1 hour before HNE, a molar ratio of Boroval to HNE of 600:1, did not ameliorate emphysema by any of our criteria. Some criteria even showed worsening of emphysema. Intratracheal administration of 250 µg of HNE, premixed with and inactivated by Boroval in 1:40 molar ratio, resulted in production of emphysema comparable in severity to that produced by 250 µg of HNE alone.

The administration of 500 µg of HNE premixed with 1:40 molar ratio of Boroval resulted in emphysema of a severity not previously seen with HNE - the mean linear

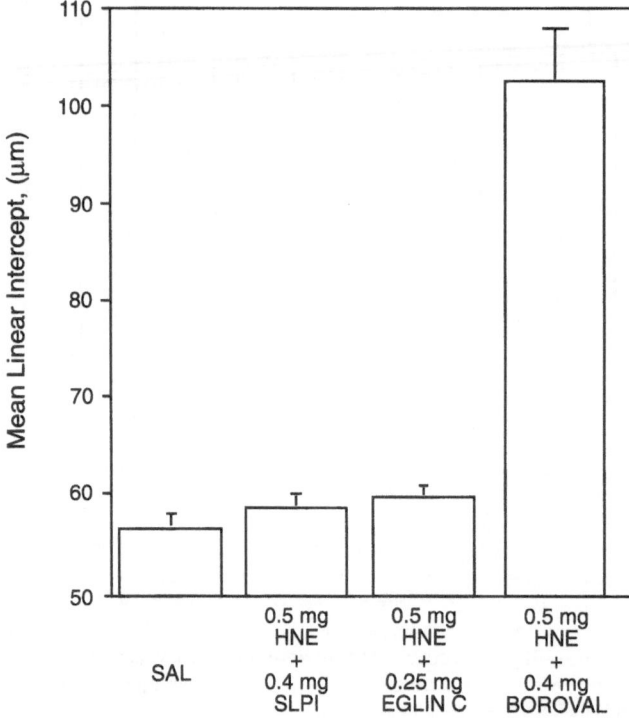

Fig. 2. Effect of complexes of 0.5 mg human neutrophil elastase (HNE) with 0.4 mg secretory leucocyte protease inhibitor (SLPI), 0.25 mg eglin-c, or 0.4 mg methoxy succinyl-L-alanyl-L-alanyl-L-prolyl-ambo-boroval pinacol (Boroval) on respiratory airspace size in hamsters as estimated by the mean linear intercept (MLI). Note that after intratracheal treatment with the HNE+SLPI and HNE+eglin-c complexes, the MLI is comparable to the saline group. The MLI after the HNE+Boroval complex is almost double that of the saline control. A positive control is not shown because >0.3 mg of HNE consistently kills all the hamsters; 0.3 mg HNE produces about a 20% increase in MLI.

intercept was almost double the negative control instead of the approximately 20% increase usually achieved with 250-300 µg of HNE (Fig. 2).

Note that doses of HNE >300 µg, which are not complexed with inhibitor, have consistently killed hamsters within 2 hours of instillation by causing massive pulmonary haemorrhage. In contrast, 500 µg of HNE premixed with a 40-fold molar excess of Boroval produced neither pulmonary haemorrhage nor deaths.

We hypothesized that the transient inactivation of HNE by Boroval largely prevented the destructive effects of HNE on the alveolar walls and their capillaries. Alveolar haemorrhage and the associated heavy influx of antiproteases, which partially inactivate the HNE, did not occur. At the same time, the HNE-Boroval complex was transported across the alveolar epithelium into the interstitium of the alveolar wall. Once in the interstitium, the complex dissociated and the Boroval cleared; the HNE proceeded to digest the elastin, thus giving rise to emphysema. It is as if the Boroval, acting like a Trojan horse, provided safe passage for the HNE into the alveolar wall.

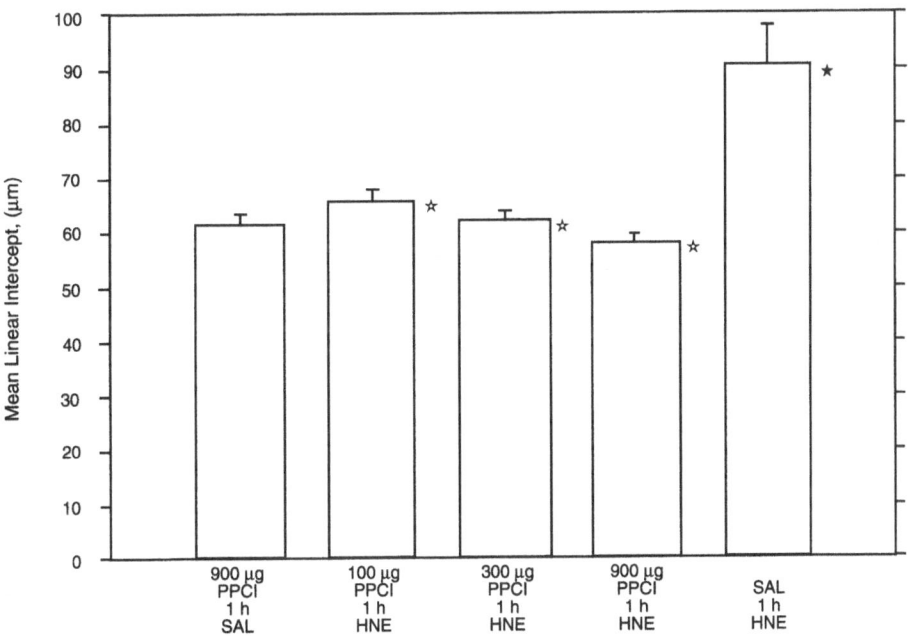

Fig. 3. Effects of various doses of polymer-bound peptidyl carbamate inhibitor (PPCI) in protecting against emphysema induction by human neutrophil elastase (HNE). Groups of hamsters received 100, 300 or 900 µg of PPCI and 1 hour later, were given 300 µg of HNE. At 56 days, the mean linear intercept of the lungs, a measure of airspace size, was determined. The positive control (saline followed in 1 hour by HNE) was the only group that was significantly different (solid star, $p<0.05$) from the negative control group (900µg PPCI followed in 1 hour by saline). All PPCI groups were significantly different from the positive control group (open star, $p<0.05$). There was amelioration of emphysema with all doses of PPCI.

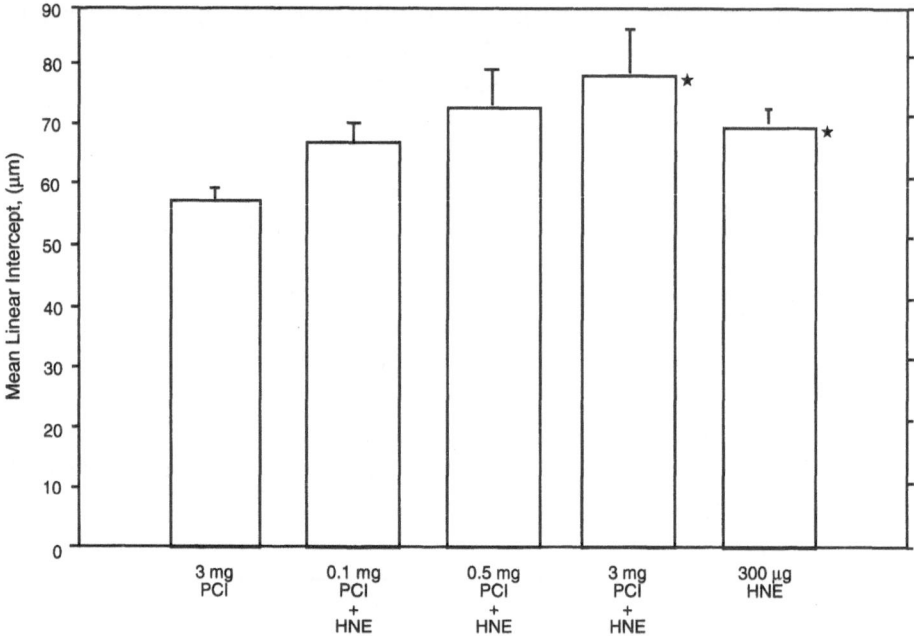

Fig. 4. Mean linear intercept values (mean±SE) for lungs of hamsters given a single intratracheal dose of peptidyl carbamate (PCI) or a mixture of human neutrophil elastase (HNE) and PCI. At 56 days, the mean linear intercept of the lungs, a measure of airspace size, was determined. The positive control group given 300 µg HNE was significantly different (solid star, p<0.05) from the negative control group (3.0 mg PCI in saline). The mean linear intercepts of all groups receiving the complex of PCI and HNE were larger than the negative control group, but the 3 mg PCI group was the only one that was significantly different from the negative control. Although not significantly different from it, all PCI groups were equal to or larger than the positive control group, indicating that amelioration of emphysema did not occur.

A recent study of a novel peptidyl carbamate inhibitor, methoxysuccinyl-L-alanyl-L-alanyl-L-propyl-CH_2N (i-Pr)CO_2-p-nitrophenol (PCI)[33] studied alone and bound to a water-soluble polymer, N(2-hydroxyethyl)-D,L-aspartamide, (PPCI) provides evidence in support of the Trojan horse theory. The molecular mass of the PCI is 591, that of PPCI is 22,000. The *in vitro* HNE-MR$_{50}$ of PCI was 4.4, that of PPCI was 0.38. The LCt$_{1/2}$ of the PCI was 3-6 min, that of the PPCI was much greater, 7.6 hours.

Figure 3 shows the effects of intratracheal administration to hamsters of 100, 300 or 900 µg of PPCI 1 hour before intratracheal administration of 300 µg HNE. Emphysema was virtually completely ameliorated in all groups.

Figure 4 shows the results of giving 0.1, 0.5 or 3 mg of PCI mixed with 300 µg of HNE intratracheally to hamsters; more than enough PCI was present in all instances to completely inhibit the HNE. Nevertheless, significant amelioration of emphysema did not occur with any dose of PCI and, like the positive control, the

3 mg PCI+HNE group was significantly different from the negative control. Thus, unless PCI is bound to a polymer, with a resultant marked increase in molecular weight and lung clearance time, the reversible inhibitor PCI fails to ameliorate emphysema, even when present in gross excess as compared with HNE.

Determinants of in-vivo *efficacy of inhibitors*

The interaction among three factors may be considered to determine the behaviour of a soluble elastase inhibitor in preventing HNE-induced emphysema: the rate of dissociation of the HNE-inhibitor complex, the rate of clearance of the inhibitor from the lungs and the rate of transport of the HNE-inhibitor complex across the alveolar epithelium. As already noted, irreversible and tightly binding inhibitors are effective in preventing emphysema induction by HNE. Two reversible inhibitors with similar dissociation rate constants may have very different effects if their rates of clearance from the lungs are different. Slow clearance of the inhibitor from the lungs favours recomplexing with HNE and protection against emphysema. Slow transport of the complex into the interstitium and elastic fibre damage can occur.

On the other hand rapid clearance from the lungs of the inhibitor, and rapid transport of the complex into the interstitium of the lungs will, by the Trojan horse theory, favour the development of emphysema. In both cases intra-alveolar haemorrhage is minimized for reasons that are not currently understood.

The significance of this classification in drug development seems obvious enough except for Category 4 agents. The Trojan horse theory of potentiation of emphysema by small molecular weight reversible elastase inhibitors is not proven. Even if it were true, we do not know whether the theory might apply to HNE already in the lungs of a human smoker under treatment with a reversible inhibitor. What is clear from our Boroval[30] and PCI/PPCI[34] experiments is that the protection by an inhibitor against HNE-induced pulmonary haemorrhage in an animal must not be equated with protection against HNE-induced emphysema.[36] *In vivo* experiments on the protective effects of elastase inhibitors on HNE-induced emphysema should be part of the preclinical process of developing HNE inhibitors for use as drugs for the preventive treatment of emphysema.

Testing of Antielastases in Humans

A discussion of the toxicologic and pharmacologic studies which must be done prior to efficacy testing is beyond the scope of this article. So is a discussion of the pros and cons of administration of a drug by the aerosol route as compared with the oral route. Testing the therapeutic effect of antielastase drugs in people raises two questions: what population should be selected for study and how should the population be studied?

The study population

The structural changes of emphysema are not reversible, as far as we know. Consequently, antielastase drugs must be evaluated in smokers with the early stages of emphysema who cannot stop smoking. Since only 15% of smokers develop chronic airflow obstruction we must have some way of identifying the at-risk smoker. Longitudinal studies show that in long-time smokers, there is a strong association between the initial forced expiratory volume for 1 second (FEV_1) and the rate of fall of FEV_1 over time.[1,2,4,37] There is also evidence indicating that emphysema is the major lesion in smokers with severe airflow obstruction.[38] Thus it seems reasonable to identify chronic smokers whose FEV_1 is one or more litres below their predicted value, as at-risk for the development of emphysema. The control group in these studies should be smokers without airflow obstruction who are matched for age, sex and smoking intensity with the high risk group.

Methods of study

The efficacy of antielastase treatment in preventing the progression of emphysema in susceptible smokers can be determined by a change in outcome as shown by a decrease in the rate of decline of the FEV_1, or by a change in the process imputed to cause emphysema.

Study of outcome

A study of outcome will be difficult and expensive, as shown by Burrow's analysis of the difficulties attending an outcome study of α-1-PI replacement therapy in homozygous α-1-PI deficiency.[39] The parameters of a currently ongoing intervention trial in subjects with chronic obstructive pulmonary disease are even more forbidding. The National Heart Lung and Blood Institute (NHLBI) Lung Health Study, is comparing the longitudinal effects on lung function of two interventions: smoking cessation alone, and smoking cessation with bronchodilator therapy, with usual treatment serving as the control group. There are 2000 subjects in each arm of the study, or 6000 subjects, collected by 10 centres; the subjects will be followed for 5 years.

The costs of this study have strained the resources of the NHLBI. Certainly such an expensive and difficult study will never be done with an antielastase agent until there is some evidence in humans of its possible efficacy.

There is much evidence that proteases in the bronchial secretions of CF patients may play a role in the pathogenesis of their disease.[40-42] In a preliminary, double-blind, crossover trial comparing 4 times daily aerosolized amiloride (a sodium channel blocker) with vehicle alone, the mean loss of FVC on amiloride was less than half the loss on vehicle; there was no effect on FEV_1.[43] It is rather surprising that a change in rate of loss of lung function occurred in such a short time in this study. A positive result on lung function and protection from proteolytic injury of a controlled trial of an aerosolized antiprotease in CF would provide support for such a trial in COPD.

Study of process

There are two main approaches to the study of the process believed to lead to emphysema: evaluation of the elastase load of the lungs and detection of destruction of elastin as indicated by an increase in the level of metabolites of elastin in blood or urine. The elastase load of the lungs may be defined as the product of the elastase concentration in neutrophils and the total number of neutrophils in the bronchoalveolar lavage (BAL) fluid plus the total amount of immunologically measured elastase in BAL supernatant. It is well known that BAL is not a satisfactory measure of the microenvironments of the lungs - but currently there is none better. It seems reasonable to express results both in terms of BAL fluid and in terms of epithelial lining fluid, using urea measurements in plasma and BAL fluid to calculate the latter.[44] The urea method has shortcomings that can be minimized by using a short dwell time for the BAL.[45] Improved methods of estimating epithelial lining fluid concentrations, using [99m]Technetium as a marker, are under development.[46]

The recent work of Weitz[47] showing an increase in elastase-derived fibrinopeptides in the blood of smokers is evidence of *in vivo* elastolysis but does not prove occurrence of the process in the lungs, nor does it prove that elastin damage has occurred. However, if an antielastase were being given by aerosol, this approach might be used to indicate the effect of the antielastase on decreasing the amount of neutrophil elastase passing from the lungs into the blood. Another group of investigators has recently indicated that they were unable to reproduce Weitz's results.[48]

It has been shown in animal studies that when lung elastin is damaged by instillation of elastase into the lungs, the desmosine cross-links of elastin appear intact in the urine.[49] An increase of elastin-derived peptides has also been shown in the blood of smokers with emphysema as compared with control smokers using antibodies to these peptides.[50] Elastin peptides have also been measured in urine by Davidson and colleagues,[51] yielding values about 10-fold higher than plasma values in the same persons; plasma and urine yielded similar differences between normals, smokers without airflow obstruction and smokers with airflow obstruction. The last group yielded values 2-3 fold higher than in the two control groups, which were similar. None of these methods give any indication of the tissue source of the damaged elastin.

Investigators have not reported elevated levels of desmosine in the urine of smokers with airflow obstruction as compared with normal smokers or nonsmokers.[52,53] Urinary desmosine values were not higher in PiZZ subjects with emphysema or PiZZ children than in normals.[54]

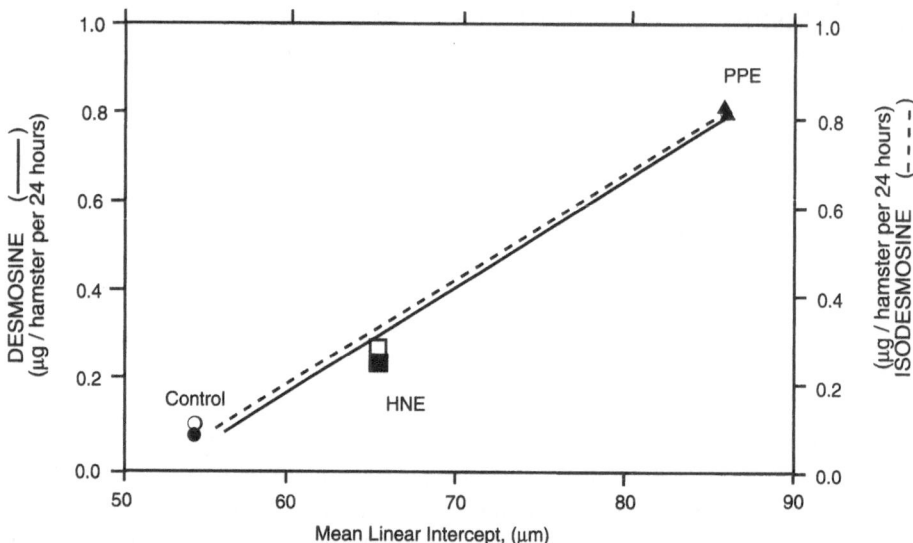

Fig. 5. Relation between urinary desmosine and isodesmosine values and emphysema after elastase treatment. Hamsters were treated intratracheally with 300 μg human neutrophil elastase (HNE, squares) or 300 μg porcine pancreatic elastase (PPE, triangles). Control animals (circles) received saline. Urine was collected for 3 days, bronchoalveolar lavage was performed, and airspace enlargement was estimated by measuring the mean linear intercept on histologic sections. Note the linear relation between desmosine (solid symbols, solid line) isodesmosine (open symbols, broken line) and the mean linear intercept. Note also the comparability of desmosine and isodesmosine values.

Stone has recently reported a new method[35] for desmosine analysis in which the urine is spiked with radiolabelled desmosine; isotope dilution is used to control for losses during purification by column chromatography, and HPLC is then used to measure desmosine. By this method, 24 hour urinary desmosine values are 10-20% of previously reported values in both animals and humans. For example, Harel et al.[53] reported adult, human, 24 hour urinary desmosine values of 47±3μg (n=23); Davies et al.[52] reported values of 75-150μg (n=157); and Bruce et al.[42] reported values of 89±11μg (n=10). Stone found this value to be 10.6±1.2μg (n=7). Kuhn et al.[55] reported urinary desmosine values in hamsters of 1.58μg/24h as compared with values of 0.074±0.008μg (n=8) obtained by Stone.

Figure 5 shows data from hamsters treated with either 300μg of porcine pancreatic elastase (PPE) or 300μg of HNE; there is a linear relation between desmosine and isodesmosine in the urine and the severity of emphysema.

A possible explanation of the failure of previous investigators to show a relation between smoking or emphysema and desmosine in the urine, is that the methods used measured both desmosine and interfering substances in the urine. Pyridinoline, a collagen cross-link derived from hydroxylysine, with a similar structure to desmosine, has been identified as one such possible substance; there may be others.[56,57] If up to 90% of previously recorded desmosine values were impurities, a doubling of desmosine output would increase the measured value by only 10-20%, an amount well within the experimental error. Thus, it is possible that the reported absence of increased desmosine in the urine of smokers and persons with emphysema is a spurious result due to a methodologic error.

The plasma and urinary elastin peptide methods have shown significant differences between groups of smokers and non-smokers, although with overlap between groups. Concordance of changes after antielastase treatment of lung elastase load, and blood and urine elastin degradation products, would give confidence that a drug effect had occurred. Although more work remains to be done, the tools necessary for a study of the efficacy of an antielastase in influencing the process believed to lead to emphysema in smokers are rapidly being put into place.

Supported by the Veterans Health Services and Research Administration, Department of Veterans Affairs, and the National Heart Lung and Blood Institute, Program Project Grant No. HL 19717

References

1. Higgins M.W., Keller J.B., Becker M., Landis J.R., Rotman H., Weg J.G., Higgins I.: An index of risk for obstructive airways disease. Am. Rev. Respir. Dis. 1982; 116: 403-410
2. Higgins M.: Epidemiology of COPD. Chest 1984; 85: 3s-8s
3. U.S. Department of Health and Human Services. Chronic obstructive lung disease. The health consequences of smoking. A report of the Surgeon General. 1984 (PHS publication #84-50205)
4. Buist S.A.: Smoking and other risk factors. In: Murray J.F., Nadel J.A. (Eds) *Textbook of Respiratory Medicine* Philadelphia, W.B. Saunders Co. 1988; 1001-1029
5. *Statistical compendium on adult lung disease*, 1987. Epidemiology and Statistics Unit, Division of Medical Affairs. American Lung Association, New York, NY
6. Progress in chronic disease prevention. *Cigarette smoking in the United States*, 1986; MMWR 1987; 36: 561-585
7. Masironi R., Rothwell K.: Tendences et effects du tabagisme dans le monde. World Health Statistics Quarterly 1988; 41: 228-239
8. Fischer E.B., Rost K.: Smoking cessation: a practical guide for the physician. Clin. Chest. Med. 1986; 7: 551-566
9. Janoff A.: Elastase and emphysema. Current assessment of the protease-antiprotease hypothesis. Am. Rev. Respir. Dis. 1985; 132: 417-433
10. Snider G.L., Lucey E.C., Stone P.J.: Animal models of emphysema. Am. Rev. Respir. Dis. 1986; 133: 149-169
11. Stockley R.A.: Alpha-1-antitrypsin and the pathogenesis of emphysema. Lung 1987; 165: 61-77
12. Senior R.M., Connolly N.L., Cury J.D., Welgus H.G., Campbell E.J.: Elastin degradation by human alveolar macrophages: a prominent role of metalloproteinase activity. Am: Rev. Respir. Dis. 1989; 139:1251-1256
13. Chapman H.A., Reilly J.J., Kobzik L.: Role of plasminogen activator in degradation of extracellular matrix protein by live human alveolar macrophages. Am. Rev. Respir. Dis. 1988; 137: 412-419

14. Willems L.N.A., Otto-Verberne C.J.M., Kramps J.A., tenHave-Oproek A.A.W., Dijkman J.H.: Detection of antileukoprotease in connective tissue of the lung. Histochemistry 1986; 86: 165-168

15. Hubbard R.C., Crystal R.G.: Antiproteases and antioxidants: strategies for the pharmacologic prevention of lung destruction: Respiration (Supplement) 1986; 50: 56-73

16. Junod A.F.: Data on oxidants and antioxidants. Bull. Eur. Physiolpathol.Respir. 1986; 22:253s-255s

17. Theron A., Anderson R.: Investigation of the protective effects of the antioxidants ascorbate, cysteine, and dapsone on the phagocyte-mediated oxidant inactivation of human alpha-1-protease inhibitor in vitro. Am. Rev. Respir. Dis. 1985; 132: 1045-1054

18. Gadek J.E., Kelman J.A., Fells G., Weinberger S.E., Horwitz A.L., Reynolds H.Y., Fulmer J.D., Crystal R.G.: Collagenase in the lower respiratory tract of patients with idiopathic pulmonary fibrosis. N. Engl. J. Med. 1979; 301: 737-742

19. Carp H., Miller F., Hoidal J., Janoff A.: Alpha$_1$-proteinase inhibitor purified from lungs of cigarette smokers contains oxidized methionine and has decreased elastase inhibitory capacity. Proc. Natl. Acad. Sci. USA 1982; 779: 2041-2045

20. Stone P.J., Calore J.D., McGowan S.E., Bernardo J., Snider G.L., Feanzblau C.: Functional apha-1-protease inhibitor in the lower respiratory tract of cigarette smokers is not decreased. Science 1983; 221:1187-1189

21. Boudier C., Pelletier A., Pauli G., Beith J.G.: The functional activity of alpha$_1$-proteinase inhibitor in bronchoalveolar lavage fluids from healthy human smokers and nonsmokers. Clin. Chim. Acta 1983; 332: 309-315

22. Abboud R.T., Fera T., Ritechter A., Tabon M.Z., Johal S.: Acute effect of smoking on the functional activity of alpha$_1$-proteinase inhibitor in bronchoalveolar lavage fluid. Am. Rev. Respir. Dis. 1985; 131: 79-85

23. Stevens M.D., Miller E.J., Cohen A.B.: Search for drugs that may reduce the load of neutrophilic azurophilic granule enzymes in the lungs of patients with emphysema. Exp. Lung Res. 1989; 15: 663-680

24. Cohen A.B., Girard W., McLarty J., Starcher B., Stevens M., Fair D.S., James H., Rosenbloom J., Kucich J.: A controlled trial of colchicine to reduce the elastase load in the lungs of cigarette smokers with chronic obstructive pulmonary disease. Am. Rev. Respir. Dis. 1990; 142: 63-72

25. Dunlap R.P., Stone P.J., Abeles R.H.: Reversible, slow, tight-binding inhibition of human leukocyte elastase. Biochem. Biophys. Res. Commun. 1987; 145: 509-513

26. Snider G.L., Stone P.J., Lucey E.C., Breuer R., Calore J.D., Seshadri T., Catanese A., Maschler R:, Schnebli H.P.: Eglin-c, a polypeptide derived from the medicinal leech, prevents human neutrophil elastase-induced emphysema and bronchial secretory cell metaplasia in the hamster. Am. Rev. Respir. Dis. 1985; 132: 1155-1161

27. Lucey E.C., Stone P.J., Christensen T.G., Breuer R., Calore J.D., Snider G.L.: Effect of varying the time interval between intratracheal administration of eglin-c and human neutrophil elastase on prevention of emphysema and secretory cell metaplasia in hamsters: with observations on the fate of eglin-c and the effect of repeated instillations. Am. Rev. Respir. Dis. 1986; 134: 471-475

28. Stone P.J., Lucey E.C., Virca G.D., Christensen T.G., Breuer R., Snider G.L.: Alpha$_1$-protease inhibitor moderates human neutrophil elastase-induced emphysema and secretory cell metaplasia in hamsters. Eur. Respir. J. 1990; 3: 673-678

29. Schanker L.S.: Drug absorption from the lung. Biochem. Pharmacol. 1978; 27: 381-385

30. Stone P.J., Lucey E.C., Snider G.L.: Induction and exacerbation of emphysema in hamsters with human neutrophil elastase inactivated reversibly by a peptide boronic acid. Am. Rev. Respir. Dis. 1990; 141:47-52

31. Kettner C.A., Shenvi A.B.: Inhibition of the serine proteases leukocyte elastase, pancreatic elastase, cathepsin G., and chymotrypsin by peptide boronic acids. J. Biol. Chem. 1984; 259: 15106-15114

32. Soskel N.T., Watanabe S., Hardie R., Shenvi A.B., Punt J.A:, Kettner C.: A new peptide boronic acid inhibitor of elastase-induced lung injury in hamsters. Am. Rev. Respir. Dis. 1986; 133: 639-642

33. Digenis G.A., Agha B.J., Tsuji K., Kato M., Shinogi M.: Peptidyl carbamates incorporating amino acid isosteres as novel elastase inhibitors. J. Med. Chem. 1986; 29: 1468-1476

34. Lucey E.C., Stone P.J., Digenis G.A., Snider G.L.: A polymer bound elastase inhibitor is effective in preventing human neutrophil elastase-induced emphysema. Ann. NY Acad. Sci.1991; 624: 341-342

35. Stone P.J., Lucey E.C., Bryan-Rhadfi J., Snider G.L., Franzblau C.: Isolation of urinary desmosine by HPLC, amino acid analysis and quantification by isotope dilution. Ann. NY Acad. Sci. 1991; 624: 355-357

36. Bond J.A., Wolff R.K., Harkema J.R., Mauderly J.L., Henderson R.F., Griffith W.C., McClellan R.O.: Distribution of DNA adducts in the respiratory tract of rats exposed to diesel exhaust. Toxicol Appl. Pharmacol 1988; 96:336-346

37. Fletcher C.R., Peto R., Tinker C., Speizer F.E.: *The Natural History of Chronic Bronchitis and Emphysema.* New York, Oxford University Press, 1976

38. Snider G.L.: Chronic obstructive pulmonary disease: risk factors, pathophysiology and pathogenesis. Ann. Rev. Med. 1989;40:411-429

39. Burrows B.: A clinical trial of efficacy of antiproteolytic therapy: can it be done? Am. Rev. Respir. Dis. 1983;127:s42-s43

40. Berger M., Sorensen R.U., Tosi M.F., Dearborn D.G., Doring G.: Complement receptor expression on neutrophils at an inflammatory site, the Pseudomonas-infected lung in cystic fibrosis. J. Clin. Invest. 1989; 84:1304-1313

41. Suter S., Schaad U.B., Morganthaler J.J., Chavallier L., Schnebli H.P.: Fibronectin-cleaving activity in bronchial secretions of patients with cystic fibrosis. J. Infect. Dis. 1988;158:89-100

42. Bruce M.C., Poncz L., Klinger J.D., Stern R.C., Tomashefski J.F., Dearborn D.G.: Biochemical and pathologic evidence for proteolytic destruction of lung connective tissue in cystic fibrosis. Am. Rev. Respir. Dis. 1985;132:529-535

43. Knowles R.M., Church N.L., Waltner W.E., Yankaskas J.R., Gilligan P., King M., Edwards L.J., Helms R.W., Boucher R.C.: A pilot study of aerosolized amiloride for the treatment of lung disease in cystic fibrosis. N. Engl. J. Med. 1990; 322:1189-1194

44. Rennard S., Basset G., Lecossier D., O'Donnel K., Martin P., Crystal R.G.: Estimation of volume of epithelial lining fluid recovered by lavage using urea as a marker of dilution. J. Appl. Physiol. 1986;60:532-538

45. Marcy T.W., Merril W.W., Rankin J.A., Reynolds H.Y.: Limitations of using urea to quantify epithelial lining fluid recovered by bronchoalveolar lavage. Am. Rev. Respir. Dis. 1987;135:1276-1280

46. Peterson B.T., Idell S., MacArthur C., Cohen A.B.: Modification of the bronchoalveolar lavage procedure to allow measurement of solute concentrations in lung epithelial lining fluid. Am. Rev. Respir. Dis. 1988; 137:147

47. Weitz J.I., Crowley K.A., Landman S.L., Lipman B.I., Yu J.: Increased neutrophil elastase activity in cigarette smokers. Ann. Int. Med. 1987;107:680-682

48. Mumford R.A., Williams M., Mao J., Dahlgren M.E., Frankenfield D., Nolan T. et al.: Direct assay of $A\alpha$ (1-21), a PMN elastase specific cleavage product of fibrinogen, in the chimpanzee.

158

Ann. NY Acad. Sci. 1991; 624: 167-178

49. Goldstein R.A., Starcher B.C.: Urinary excretion of elastin peptides containing desmosine after intratracheal injection of elastase in hamsters. J. Clin. Invest. 1978; 61:1286-1290

50. Kucich U., Christner P., Lippmann M., Kimbel P., Williams G., Rosenbloom J., Weinbaum G.: Utilization of a peroxidase antiperoxidase complex in an enzyme-linked immunosorbent assay of elastin-derived peptides in human plasma. Am. Rev. Respir. Dis. 1985;131:709-713

51. Schriver E.E., Bernard G.R., Swindell B.B., Sutcliffe M.S., Davidson J.M.: Elastin fragment levels in human plasma, urine and bronchoalveolar lavage fluid (BALF). Chest 1989; 96:153s

52. Davies S.F., Offord K.P., Brown M.G., Campe H., Niewoehner D.: Urine desmosine is unrelated to cigarette smoking or to spirometric function. Am. Rev. Respir. Dis. 1983; 128:473-475

53. Harel S., Janoff A., Yu S.Y., Hurewitz A., Bergofsky E.H.: Desmosine radioimmunoassay for measuring elastin degradation in vitro. Am. Rev. Respir. Dis. 1980;122:769-773

54. Pelham F., Wewers M., Crystal R., Buist A.S., Janoff A.: Urinary excretion of desmosine (elastase cross-links) in subjects with PiZZ alpha1-antitrypsin deficiency, a phenotype associated with hereditary predisposition to pulmonary emphysema. Am. Rev. Respir. Dis. 1985;132:821-823

55. Kuhn C., Engleman W., Chraplyvy M., Starcher B.C.: Degradation of elastin in experimental elastase-induced emphysema measured by a radioimmunoassay for desmosine. Exp. Lung. Res. 1983;5:115-123

56. Gunga-Smith Z.: An emzyme-linked immunosorbent assay to quantitate the elastin crosslink desmosine in tissue and urine samples. Anal. Biochem. 1985;147:258-264

57.Laurent P., Magne L., DePalmas J., Bignon J., Jaurand M.C.: Quantitation of elastin in human urine and rat pleural mesothelial cell matrix by a sensitive avidin-biotin ELISA for desmosine. J. Immunologic Methods 1988; 107:1-11

58. Lively M.O., Powers J.C.: Specificity and reactivity of human granolocyte elastase and cathepsin G, porcine pancreatic elastase, bovine chymotrypsin and trypsin toward inhibition with solfonyl fluorides. Biochim. Biophys. Acta 1978; 525:171-179

59. Zimmerman M., Morman H.J., Nulvey D., Jones H., Frankshun R., Ashe B.M.: Inhibition of elastase and other serine proteases by heterocyclic acylating agents. J. Biol. Chem. 1980;255:9848-9851

60. Harper J.W., Hemmi K., Powers J.C.: Reaction of serine proteases with substituted isocoumarins: discovery of 3,4-dichloroisocoumarin, a new general mechanism based serine protease inhibitor. Biochemistry 1985; 24: 1831-1841

61. Hori H., Yasutake A., Minematsu Y., Powers J.C.: Inhibition of human leukocyte elastase, porcine pancreatic elastase and cathepsin G by peptide ketones. In: Deber C.M., Hruby V.J., Dopple K.D. (Eds.) *Peptides: Structure and Function, Proceedings of Ninth American Peptide Symposium.* Rockford IL, Pierce Chemical Co. 1985; 819-822

62. Lucey E.C., Stone P.J., Powers J.C., Snider G.L.: Amelioration of human neutrophil elastase-induced emphysema in hamsters by pretreatment with an oligopeptide chloromethyl ketone. Eur. Respir. J. 1989; 2:421-427

63. Lucey E.C., Stone P.J., Ciccolella D.E., Breuer R., Christensen T.G., Thompson R.C., Snider G.L.: Recombinant human secretory leukocyte inhibitor: in vitro properties and amelioration of human neutrophil elastase-induced emphysema and secretory cell metaplasia in the hamster. J. Lab. Clin. Med. 1990;115:224-232

12. Genetic Control of Human Alpha-1-Antitrypsin and Hepatic Gene Therapy

S. L. C. Woo[1,2], R. N. Sifers[2,3], K. Ponder[2]

1. Howard Hughes Medical Institute, Baylor College of Medicine, Houston, Texas, USA
2. Department of Cell Biology, Baylor College of Medicine, Houston, Texas, USA
3. Department of Pathology, Baylor College of Medicine, Houston, Texas, USA

Introduction

Alpha-1-antitrypsin (AAT) is a 50 kDa serine protease inhibitor which is synthesized most abundantly by hepatocytes.[1,2] It accounts for approximately 90% of the total protease inhibitory capacity in normal human serum.[3,4] In common with other members of the serine protease inhibitor superfamily,[5] the active inhibitory site of AAT is centred[6] around Ser[358]. Although it exhibits the capacity to inhibit a variety of proteases, including trypsin, chymotrypsin, thrombin, kallikrein, and plasmin[7], its apparent major physiological role is to diffuse into the alveoli structure of the lung, and protect elastin fibres from excessive hydrolysis by neutrophil elastase.[8,9] Neutrophil elastase is inhibited by forming a pseudo-irreversible equimolar complex of the protease with inhibitor.[10]

Under normal physiological conditions, serum AAT levels range from 150 to 350 mg/dl, which is sufficient to provide a protease/antiprotease balance for the protection of lung tissues.[11] However, AAT is extremely polymorphic and several variants exist which are associated with either a partial deficiency or total absence of the protease inhibitor in sera.[11] Inheritance of specific combinations of these variant alleles can result in a predisposition toward the development of pulmonary emphysema, and in some cases, liver cirrhosis.

Molecular and Genetic Basis of α-1-Antitrypsin Deficiency

The initial molecular analysis of this genetic disease included the mapping of the AAT structural gene locus to the q31-32 region of human chromosome 14[12,13] which has been designated Pi. At present, over 75 variants of AAT have been identified which are allelic and codominantly expressed.[14] The most common AAT allele is designated PiM and exhibits an overall gene frequency of 0.95.[11] PiM alleles have been subclassed as M1, M2, and M3[15,16] and inheritance of any or a combination of these alleles will result in sufficient serum AAT levels to protect lung tissues from excessive degradation by neutrophil elastase.

The two most frequent variants associated with a deficiency of the circulating level of AAT are designated PiS and PiZ, which exhibit allelic frequencies of 0.02-0.03 and 0.01-0.02, respectively.[17,15] Our initial molecular characterization of the PiS allele revealed that an A to T transversion corresponding to a Glu to Val substitution at residue 264 in the protein is responsible for its deficiency phenotype.[18] The subsequent characterization of the PiZ allele by us and others[19,20] revealed the presence of a T to C transition in exon III and a G to A transition in exon V of this allele resulting in Val[213] to Ala[213], and Glu[342] to Lys[342] substitutions, respectively. Ala[213] is also present in the normal PiM1 allele, and not related to the deficiency syndrome.[16] Subsequent expression studies[20] have demonstrated that the Glu[342] to Lys[342] substitution is solely responsible for the deficiency phenotype associated with the PiZ allele. Oligonucleotides that specifically hybridize with this mutant sequence have been synthesized[21] and used successfully for the prenatal diagnosis of the PiZ AAT genotype.[22]

Additional rare AAT alleles which are associated with either a partial deficiency[23-,25] or total absence of serum AAT[26-29] have been identified. In some cases the molecular defect has been characterized.[1,16,30-33] Although various serum AAT levels occur in individuals carrying specific combinations of these alleles, it is the severe deficiency or total absence of this protease inhibitor in sera which predisposes individuals toward the development of pulmonary emphysema.[11]

Biochemical Basis of α-1-Antitrypsin Deficiency

The dramatic polymorphism of the AAT macromolecule is indicative that a variety of mechanisms might lead to the various deficiency and "null" phenotypes. For instance, although a G to A transition results in a Glu to Val substitution in the PiS variant[18], the major cause of its decreased levels in sera most likely results from the presence of an aberrant splice site introduced by the point mutation, and results in the improper processing of a fraction of the mRNA transcripts. Alternatively, the Glu to Lys substitution in the PiZ variant causes it to aggregate within the lumen of the rough endoplasmic reticulum (RER), such that only a small fraction of the protein is secreted into the bloodstream.[34,35]

Identical to other hepatic secretory proteins, AAT is translocated into the lumen of the RER during its biosynthesis. Recent studies have demonstrated that several genetically-engineered and naturally occurring mutant proteins are retained within the RER following their synthesis[36-39], suggesting that proteins must first achieve a proper structure of conformation prior to their export from this subcellular compartment.[40] Because AAT must transit the RER prior to secretion into the bloodstream, it is conceivable that any of a variety of mutations might result in the retention of a mutant AAT macromolecule within this subcellular compartment. This hypothesis was recently verified by our observation that the Pi Null$_{\text{Hong Kong}}$ AAT variant is retained in the RER of stably transfected hepatoma cells.[1] Furthermore, two AAT deficiency phenotypes designated PiM$_{\text{Malton}}$ and PiM$_{\text{Duarte}}$ are both associated with the presence of PAS-positive inclusion bodies in hepatocytes[23,25], suggesting that an intracellular transport defect of these variant proteins is the cause of the deficiency phenotype. Furthermore, post-transcriptional intracellular degradation is the likely mechanism for the phenotypes associated with the M$_{\text{Heerlen}}$ M$_{\text{Procida}}$ and Null$_{\text{Mattawa}}$ AAT variants.[30,31,33] Conceivably, degradation of these mutant proteins occurs shortly after their synthesis and translocation into the lumen of the RER.[41]

Understandably, a major area of interest in modern cell biology is the elucidation of the molecular mechanisms that organize and regulate the extensive protein traffic that transits through the first subcellular compartment of the exocytic pathway, the RER. The regulatory mechanisms involved in this process include the selective retention and targeted degradation of mutant proteins in this organelle. Because of the association of defective intracellular transport and/or intracellular degradation of several types of AAT mutants, an understanding of the molecular mechanism(s) associated with the regulation of protein traffic in the RER is of great importance.

Loebermann et al.[42] have reported that Glu^{342} forms an intramolecular salt bridge with Lys^{290} in the normal AAT macromolecule. Suspecting that the disruption of this structural feature by the Glu^{342} to Lys^{342} substitution in the PiZ variant might be responsible for its hindered secretion, several laboratories have analysed the secretion of mutant human AATs bearing specific amino acid substitutions at residues 290 and 342. Results from most laboratories have demonstrated that although the integrity of the 290-342 salt bridge does have a minor effect on the secretion of AAT, it is the substitution of the positively charged Lys^{342} residue that is responsible for the hindered secretion of the PiZ protein.[20,43,44] Conceivably, the introduction of the positively charged Lys residue alters the local electrostatic charge of that region of the macromolecule, thereby interfering with its proper folding during translation. Although Brantly et al.[45] have suggested that the 290-342 salt bridge plays a significant role in the proper intracellular transport of AAT, their findings have not been corroborated. The importance of the salt bridge *versus* the importance of the Lys residue at position 342 is still under investigation.

At present, one resident RER protein, BiP (immunoglobulin heavy chain binding protein) has been implicated in the retention of proteins within the RER.[46,47]

Although McCraken et al.[44], have reported that the accumulated PiZ protein is not associated with BiP, further studies must be performed to determine whether there is a transient association between these two proteins which may lead to the accumulation phenomenon.

Intrahepatic Accumulation of the Insoluble PiZ Variant Causes Liver Damage

The accumulation of aggregated human PiZ AAT within the hepatic RER results in the dramatic distension of a subset of RER cisternae.[48] Interestingly, an increased risk for the development of juvenile cirrhosis exists in some infants bearing the PiZZ phenotype.[49] Furthermore, numerous sporadic reports have associated PiZ AAT deficiency with the development of chronic liver disease in PiZZ adults.[2] At present, the potential for an increased risk for chronic liver disease in heterozygous PiMZ individuals remains controversial.

At present, two groups have utilized a transgenic mouse model to determine whether liver disease associated with PiZ AAT deficiency results from the intrahepatic accumulation of the mutant protein or whether this phenomenon is caused by the reduced circulating levels of AAT.[50,51] Cytochemical and ultrastructural analyses of liver sections from PiZ-bearing mice have demonstrated a correlation between the intrahepatic accumulation of the PiZ variant and the development of liver disease. Because these transgenic animals synthesize and secrete normal levels of their endogenous murine AAT, it has been concluded that the development of liver disease must result from the intrahepatic accumulation of the mutant PiZ protein.

Subsequent analyses have demonstrated that the accumulated PiZ protein is in the form of an insoluble aggregate[52] as it is in human subjects.[34,35] Although normal human AAT can also accumulate in livers of PiM-bearing transgenic mice due to its elevated expression[2], the accumulated protein is in a completely soluble form.[52] Since the intrahepatic accumulation of normal PiM AAT does not cause liver damage in transgenic mice[50,51] it is reasonable to conclude that the insolubility of the PiZ protein is the etiological agent of liver damage in this disorder.

Although the insolubility of the newly synthesized PiZ protein has been implicated as the direct cause of the retention of the mutant protein within the RER, we have recently demonstrated that the PiZ protein accumulates within the RER of stably transfected mouse hepatoma cells in a completely soluble form.[20] Therefore, the aggregation and subsequent insolubility of the PiZ variant cannot be the direct cause of the intrahepatic accumulation of this protein, as has been suggested.[53] Rather, the aggregation of the PiZ variant is most likely a secondary phenomenon to the actual retention mechanism. Whether the accumulated insoluble PiZ protein escapes the proteolytic degradative machinery in the RER[41], thus resulting in its accumulation, is not yet known. Furthermore, the fact that only a small percentage of PiZZ infants develop liver cirrhosis may reflect the need of additional environmental insults or the presence of auxiliary genetic determinants to result in liver damage in humans.

The Hepatic Protein Secretory Machinery Can Be Saturated

Globular inclusion bodies containing immunoreactive human AAT have been identified within hepatocytes of normal human subjects undergoing extreme acute phase situation and exhibiting an elevated synthesis of AAT.[54] Likewise, the elevated synthesis of normal human AAT within hepatocytes of PiM-bearing transgenic mice results in the accumulation of this protein within distended cisternae of the hepatic RER.[2] Furthermore, recent results from our laboratory have demonstrated that the elevated synthesis of the PiM protein in isolated hepatocytes from PiM-bearing transgenic mice results in the hindered secretion of murine AAT, but not murine transferrin or albumin which represent glycosylated and non-glycosylated hepatic secretory proteins, respectively.[52] Overall, these observations demonstrate that the elevated synthesis of human AAT can hinder the export of murine AAT from the hepatic rough endoplasmic reticulum in an apparently specific manner.

Several secretory and transmembrane proteins are transported to the cell surface at different but characteristic rates.[55,56] These observations led to the hypothesis that receptors might accelerate the export of specific proteins from the RER during the early stages of intracellular transport. However, the results of recent studies have led to the hypothesis that the normal mechanism for protein export from the RER involves a "default" pathway which is characterized by the bulk flow of proteins from this subcellular compartment into transition vesicles *en route* to the cis region of the Golgi complex.[57,58] The demonstration of an apparent specificity toward the export of human and murine AATs from the hepatic RER cannot be explained by a mere bulk flow mechanism, suggesting that additional pathways might exist for the export of proteins from the RER. Thus, future analyses utilizing AAT as a model to study the mechanisms of intracellular protein transport will undoubtedly lead to new ideas in this area of cell biology.

Correction of AAT Deficiency by Hepatic Gene Transfer

An efficient method for the transduction of functional genes into a variety of cell types using the retrovirus as a vector has recently been developed by Richard Mulligan and colleagues.[59] We and others have inserted the human AAT cDNA into such a retroviral vector and demonstrated that the chimeric virus can effectively transduce the gene into mouse fibroblasts, which were then able to synthesize *de novo* the human AAT protein and secrete it into the culture medium.[60,61]

While the virally infected fibroblasts had been reintroduced into the peritoneum cavity of rodents and shown by Crystal and colleagues that they can synthesize the human protein *in vivo*, long-term reconstitution has not yet been achieved.

Our laboratory has been investigating the possibility of reconstituting the deficiency activity by hepatic gene therapy. We were able to isolate primary hepatocytes from the transgenic mouse line expressing human AAT in liver, and demonstrated that the cells survive in culture for an extended period of time and continued to secrete human AAT into the medium. The hepatocytes were then transplanted into congenic mouse recepients, so that there would be no immune rejection of the transplanted cells. The hepatocytes were seated onto inert matrices such as collagen beads and gel foam, which were implanted into the peritoneum cavity or directly into the liver of recipient mice. Human AAT was detected in the recepient's plasma 1 day post transplantation, and the level remained relatively stable for up to 2 weeks. Thereafter, the level declined gradually and by 4 weeks, it was no longer detectable. The implants were retrieved from the recipients and examined. It was no surprise that indeed there were no live hepatocytes on the implants. Having experimented with a variety of methods to improve the survival of the transplanted hepatocytes on matrices and failed, we introduced the cells directly into the portal vein of the spleen. In these instances, the expression of human AAT in the recipient was stable for 9-12 months, which was essentially the life of the recepient mice.[62] Thus long-term reconstitution of hepatic function is feasible which could lead to the development of gene therapy for AAT deficiency in man.

References

1. Sifers R.N., Brashears-Macatee S., Kidd V.J., Muensch H., Woo S.L.C.: A frameshift mutation results in a truncated alpha-1-antitrypsin that is retained within the rough endoplasmic reticulum. J. Biol. Chem. 1988; 263: 7330-7335
2. Carlson J.A., Sifers R.N., Woo S.L.C.: Molecular defect in an inheritable liver diease: alpha-1-antitrypsin deficiency. In: Arias I.M., Jakoby W.B., Popper H., Schachter D., (Eds.). *The Liver: Biology and Pathobiology*, 2nd edition, New York Raven Press 1988; 1161-1168
3. Heimburger N., Haupt H., Schwick H.G.: Proteinase inhibitors in human plasma. In: Ischeche F.H., Berlin H., deGruyter W., (Eds.). *Proceedings of the International Research Conference on Proteinase Inhibitors*. Munich 1970; 1-22
4. Janoff A.: Inhibition of human granulocyte elastase by serum alpha-1-antitrypsin. Am. Rev. Respir. Dis. 1972; 105: 121-122
5. Hunt L.T., Dayhoff M.O.: A surprising new protein superfamily contains ovalbumin, antithrombin III, and alpha-1-proteinase inhibitor. Biochem. Biophys. Res. Commun. 1980; 95: 864-871
6. Johnson D., Travis J.: Structural evidence for methionine at the reactive site of human alpha-1-proteinase inhibitor. J. Biol. Chem. 1978; 253: 7142-7144
7. Heidtmann H., Travis J.: Human alpha-1-proteinase inhibitor. In: Barrett A.J., Salvensen G., (Eds.) *Proteinase inhibitors*. New York, Elsevier 1986; 441-456

8. Olsen G.N., Harris J.O., Castle J.R., Waldeman R.H., Karmgrad H.J.: Alpha-1-antitrypsin content in the serum, alveolar macrophages, and alveolar lavage fluid of smoking and nonsmoking normal subjects. J. Clin. Invest. 1975; 55: 427-430

9. Gadek J.E., Hunninghake G.W., Fells G.A., Zimmerman R.L., Keogh B.A., Crystal R.G.: Evaluation of the protease-antiprotease theory of human destructive lung disease. Bull. Europ. Physiopath. Resp. 1980; 16: 27-40

10. Beatty K., Bieth J.C., Travis J.: Kinetics of association of serine proteinases with native and oxidized alpha-1-proteinase inhibitor and alpha-1-antichymotrypsin. J. Biol. Chem. 1980; 255: 3931-3934

11. Gadek J.E., Crystal R.G.: Alpha-1-antitrypsin deficiency. In: Stanbury J.B., Wyngaarden J.B., Fredrickson D.S., Goldstein M.S., (Eds.). *The metabolic basis of inherited disease.* 5th Edition, Minneapolis, McGraw-Hill Publications 1982; 1450-1467

12. Schroeder W.T., Miller M.F., Woo S.L.C., Saunders G.F.: Chromosomal localization of the human alpha-1-antitrypsin gene (Pi) to 14q 31-32. Am. J. Hum. Genet. 1985; 37: 868-872

13. Rabin M., Watson M., Kidd V., Woo S.L.C., Breg W.R., Ruddle F.H.: Regional localization of alpha-1-antichymotrypsin and alpha-1-antitrypsin genes on human chromosome 14. Somat. Cell Mol. Genet. 1986; 12: 209-214

14. Brantly M., Nukiwa Y., Crystal R.G.: Molecular basis of alpha-1-antitrypsin deficiency. Am. J. Med. 1988; 84: 13-31

15. Kueppers F., Christopherson M.J.: Alpha-1-antitrypsin: further genetic heterogeneity revealed by isoelectric focusing. Am. J. Hum. Genet. 1978; 30: 359-365

16. Nukiwa T., Brantly M., Ogushi F., Fells G., Satoh K., Stier L., Courtney M., Crystal R.G.: Characterization of the M1 (ala^{213}) type of alpha-1-antitrypsin, a newly recognized, common "normal" alpha-1-antitrypsin. Biochemistry 1987; 26: 5259-5267

17. Fagerhol M.K., Cox D.W.: The Pi polymorphism. Genetic, biochemical and clinical aspects of human alpha-1-antitrypsin. Adv. Hum. Genet. 1981; 11: 1-62

18. Long G.L., Chandra T., Woo S.L.C., Davie W.W., Kurachi K.: Complete sequence of the cDNA for human alpha-1-antitrypsin and the gene for the S variant. Biochemistry 1984; 23: 4828-4837

19. Nukiwa T., Satoh K., Brantly M.L., Ogushi F., Fells G.A., Courtney M., Crystal R.G.: Identification of a second mutation in the protein coding sequence of the Z-type alpha-1-antitrypsin gene. J. Biol. Chem. 1986; 261: 15989-15994

20. Sifers R.N., Hardick C.P., Woo S.L.C.: Disruption of the 290-342 salt bridge is not responsible for the secretory defect of the PiZ alpha-1-antitrypsin variant. J. Biol. Chem. 1989; 264: 2997-3001

21. Kidd V.J., Wallace R.B., Itakura K., Woo S.L.C.: Alpha-1-antitrypsin deficiency detection by direct analysis of the mutation in the gene. Nature (London) 1983; 304: 230-234

22. Kidd V.J., Golbus M.S., Wallace R.B., Itakura K., Woo S.L.C.: Prenatal diagnosis of alpha-1-antitrypsin deficiency by direct analysis of the mutation site in the gene. N. Engl. J. Med. 1984; 310: 639-642

23. Cox D.W.: A new deficiency allele of alpha-1-antitrypsin: Pi M$_{Malton}$. In: Peeters H., (Ed.). *Protides of the Biological Fluids.* Oxford, Pergamon Press 1975; 375-378

24. Allen M.B., Ward A.M., Perks W.H.: Alpha-1-antitrypsin deficiency due to M$_{Malton}$ Z phenotype: case report and family study. Thorax 1986; 41: 568-570

25. Lieberman J., Gaidulis L., Klotz S.D.: A new deficient variant of alpha-1-antitrypsin (M$_{Duarte}$). Inability to detect the heterozygous state by alpha-1-antitrypsin phenotyping. Am. Rev. Respir. Dis. 1976; 113: 31-36

26. Talamo R.C., Landley C.E., Reed C.E., Makino S.: Alpha-1-antitrypsin deficiency. A variant with no detectable alpha-1-antitrypsin. Science 1973; 181: 70-71

27. Garver R.I., Mornex J.F., Nukiwa T., Brantly M., Courtney M., LeCocq J.P., Crystal R.G.: Alpha-1-antitrypsin deficiency and emphysema caused by homozygous inheritance of non-expressing alpha-1-antitrypsin genes. N. Engl. J. Med. 1986; 314: 762-766

28. Nukiwa T., Takahashi H., Brantly M., Courtney M., Crystal R.G.: Alpha-1-antitrypsin null$_{granite}$$_{falls}$ a nonexpressing alpha-1-antitrypsin gene associated with a frameshift to stop mutation in a coding exon. J. Biol. Chem. 1987; 262: 11999-12004

29. Muensch H., Gaidulis L., Kueppers F., So S.Y., Escano G., Kidd V.J., Woo S.L.C.: Complete absence of serum alpha-1-antitrypsin in conjunction with an apparently normal gene structure. Am. J. Hum. Genet. 1986; 38: 898-907

30. Takahashi H., Nukiwa T., Satoh K., Ogushi F., Brantly M., Fells G., Stier L., Courtney M., Crystal R.G.: Characterization of the gene and protein of the alpha-1-antitrypsin "deficiency" allele M$_{procida}$. J. Biol. Chem. 1988; 263: 15528-15534

31. Hofker M.H., Nukiwa T., van Paasen H.M.B., Nelen M., Frants R.R., Kramps J.A., Klasen E.C., Crystal R.G.: A pro->leu substitution in codon 369 of the alpha-1-antitrypsin deficiency variant Pi M$_{Heerlen}$. Am. J. Hum. Genet. 1987; 41: A 220

32. Satoh K., Nukiwa T., Brantly M.,Garver R.I.Jr., Hofker M.K., Courtney M., Crystal R.G.: Emphysema associated with a complete absence of alpha-1-antitrypsin in serum and homozygous inheritance of a stop codon in an alpha-1-antitrypsin coding exon. Am. J. Hum. Genet. 1987; 42: 77-83

33. Curiel D., Brantly M., Curiel E., Stier L., Crystal R.G.: Alpha-1-antitrypsin deficiency caused by the alpha-1-antitrypsin null$_{Mattawa}$ gene. An insertion mutation rendering the alpha-1-antitrypsin gene incapable of producing alpha-1-antitrypsin. J. Clin. Invest. 1989; 83: 1144-1152

34. Eriksson S., Larsson C.: Purification and partial characterization of PAS-positive inclusion bodies from the liver in alpha-1-antitrypsin deficiency. N. Engl. J. Med. 1975; 292: 176-180

35. Bathurst I.C., George P.M., Travis J., Carrell R.W.: Structure and functional characterization of the abnormal Z alpha-1-antitrypsin isolated from human liver. FEBS Lett. 1984; 177: 179-183

36. Lehrman M.A., Goldstein J.L., Brown M.S., Russell D.W., Schneider W.J.: Internalization-defective LDL receptors produced by genes with nonsense and frameshift mutations that truncate the cytoplasmic domain. Cell 1985; 41: 735-743

37. Rose J.K., Bergman J.E.: Altered cytoplasmic domains affect intracellular transport of the vesicular stomatitis virus glycoprotein. Cell 1983; 34: 513-524

38. Yamamoto T., Bishop R.W., Brown M.S., Goldstein J.L., Russell D.W.: Deletion of cysteine-rich region of LDL receptor impedes transport to cell surface in WHHL rabbit. Science 1986; 232: 1230-1237

39. Doyle C., Roth M.G., Sambrook J., Gething M.J.: Mutations in the cytoplasmic domain of the influenza hemagglutinin affect different stages of intracellular transport. J. Cell Biol. 1985; 100: 704-714

40. Gething M.J., McCammon K., Sambrook J.: Expression of wildtype and mutant forms of influenza hemagglutinin: the role of folding in intracellular transport. Cell 1986; 46: 939-950

41. Lippincott-Schwartz J., Bonifacino J.S., Yuan L.C., Klausner R.D.: Degradation from the endoplasmic reticulum: disposing of newly synthesized proteins. Cell 1988; 54: 209-220

42. Loebermann H., Tokuoka R., Deisenhofer J., Huber R.J.: Human alpha-1-proteinase inhibitor: crystal structure analysis of two crystal modifications, molecular and preliminary analysis of the implications for function. J. Mol. Biol. 1984; 177: 531-556

43. Foreman R.C.: Disruption of the lys-290-glu-342 salt bridge in human alpha-1-antitrypsin does not prevent its synthesis or secretion. FEBS Lett. 1987; 216: 79-82

44. McCracken A.A., Kruse K.B., Brown J.L.: Molecular basis for defective secretion of the Z variant of human alpha-1-protease inhibitor: secretion variants having altered potential for salt bridge

formation between amino acids 290 and 342. Mol. Cell Biol. 1989; 9: 1406-1414

45. Brantly M., Courtney M., Crystal R.G.: Repair of the secretion defect in the Z form of alpha-1-antitrypsin by addition of a second mutation. Science, 1989; 242: 1700-1701

46. Bole D.G., Hendershot L.M., Kearney J.F.: Posttranslational association of immunoglobulin heavy chain binding protein with nascent heavy chains in nonsecreting and secreting hybridomas. J. Cell Biol. 1986; 102: 1558-1566

47. Dorner A.J., Bole D.G., Kaufman R.J.: The relationship of N-linked glycosylation and heavy chain-binding protein associated with the secretion of glycoproteins. J. Cell Biol. 1987; 105: 2665-2674

48. Sharp H.L.: Alpha-1-antitrypsin deficiency. Hosp. Pract. 1971; 6: 83-96

49. Sharp H.L., Bridges R.A., Krivit W., Freier E.F.: Cirrhosis is associated with alpha-1-antitrypsin deficiency: a previously unrecognized inherited disorder. J. Lab. Clin. Med. 1969; 73: 934-939

50. Dycaico M.J., Grant S.G.N., Felts K., Nichols W.S., Geller S.A., Hager J.H., Pollard A.J., Kohler S.W., Short H.P., Jirik F.R., Hanahan D., Sorge J.A.: Neonatal hepatitis induced by alpha-1-antitrypsin: a transgenic mouse model. Science 1988; 242: 1409-1412

51. Carlson J.A., Rogers B.B., Sifers R.N., Finegold M.J., Clift S.M., DeMayo F.J., Bullock D.W., Woo S.L.C.: The accumulation of PiZ alpha-1-antitrypsin causes liver damage in transgenic mice. J. Clin. Invest. 1989; 83: 1183-1190

52. Sifers R.N., Rogers B.B., Hawkins H.K., Finegold M.J., Woo S.L.C.: Elevated synthesis of human alpha-1-antitrypsin hinders the secretion of murine alpha-1-antitrypsin from hepatocytes of transgenic mice. J. Biol. Chem. 1989; 264: 15696-15700

53. Cox D.W., Billingsley G.D., Callahan J.W.: Aggregation of plasma Z type alpha-1-antitrypsin suggests basic defect for the deficiency. FEBS Lett. 1986; 205: 255-260

54. Carlson J., Eriksson S., Hagerstand I.: Intra-and extracellular alpha-1-antitrypsin in liver disease with a special reference to Pi phenotype. J. Clin. Pathol. 1981; 34: 1020-1025

55. Ledford B.E., Davis D.F.: Kinetics of serum protein secretion by cultured hepatoma cells: evidence for multiple secretory pathways. J. Biol. Chem. 1983; 258: 3304-3308

56. Lodish H.F., Kong N., Snider M., Strous G.J.A.M.: Hepatoma secretory proteins migrate from the rough endoplasmic reticulum to Golgi at characteristic rates. Nature 1983; 304: 80-83

57. Rothman J.E.: Protein sorting by selective retention in the endoplasmic reticulum and Golgi stack. Cell 1987; 50: 521-522

58. Lodish H.F.: Transport of secretory and membrane glycoproteins from the rough endoplasmic reticulum to the Golgi. J. Biol. Chem. 1988; 263: 2107-2110

59. Mann R., Mulligan R.C., Baltimore D.B.: Construction of a retrovirus packaging mutant and its use to produce helper-free defective retrovirus. Cell 1983; 33: 153-159

60. Ledley F.D., Grenett H.E., Bartos D.P., Woo S.L.C.: Retroviral mediated transfer and expression of human alpha1-antitrypsin in cultured cells. Gene 1987; 61: 113-118

61. Garver R.I., Chytil A., Karlsson S., Fells G.A., Brantly M.L., Courtney M., Kantoff P.W., Neinhuis A.W., Anderson W.F., Crystal R.G.: Production of glycosylated physiologically "normal" human alpha 1-antitrypsin by mouse fibroblasts modified by insertion of a human alpha 1-antitrypsin cDNA using a retroviral vector. Proc. Natl. Acad. Sci. U.S.A. 1987; 84: 1050-1054

62. Ponder K.P., Gupta S., Leland F., Darlington G., Finegold M., DeMayo J., Ledley F.D., Chowdhury J.R., Woo S.L.C.: Mouse hepatocytes migrate to liver parenchyma and function indefinitely after intrasplenic transplantation. Proc. Natl. Acad. Sci. USA 1991; 88: 1217-1221

13. Neutrophils, Neutrophil Elastase and the Fragile Lung: The Pathogenesis and Therapeutic Strategies Relating to Lung Derangement in the Common Hereditary Lung Disorders

N. G. McElvaney, P. Birrer, L. M. Chang-Stroman, R. G. Crystal

Pulmonary Branch, National Heart, Lung and Blood Institute, National Institute of Health Bethesda, Maryland, USA

In the process of being exposed to 6000-10,000 litres of air daily, the respiratory epithelial surface is repeatedly exposed to microorganisms, inorganic and organic particulates, gases, fumes and aerosols. To help defend itself, the lung has available a variety of mechanisms that can recruit inflammatory cells, particularly blood neutrophils. However while effective in killing microorganisms, the neutrophil is a dangerous "two-edged sword", with a variety of potent mediators capable of injuring normal tissues.[1] The lung is particularly vulnerable to these neutrophil mediators, as it is very fragile, with tissues that are easily destroyed and difficult to repair.[2] The most dangerous neutrophil mediator is neutrophil elastase (NE), a 29 kDa serine protease capable of destroying most components of the extracellular matrix of the alveolar walls.[1,3] In addition, NE can injure the bronchial epithelium and interfere with respiratory host defence by altering mucociliary clearance mechanisms and interfering with the ability of inflammatory cells to kill microorganisms.[3-5]

In normal individuals, there is chronic, low level traffic of neutrophils, and hence NE, into lung tissues. However, respiratory tissues are normally protected from NE by NE-specific antiproteases, particularly α1-antitrypsin (α1AT), and to a much lesser extent, the secretory leucoprotease inhibitor (SLPI) and α2-macroglobulin (α2M) (Fig. 1; for detailed references regarding these antiproteases and their role in the lung, see ref. 6).

α1AT, a 52 kDa glycoprotein composed of a single chain of 394 amino acids with three carbohydrate side chains provides 90% of the anti-NE protection of the lower respiratory tract.[6-10]

Most of the α1AT in the lung is produced in the liver and secreted into the plasma,

170

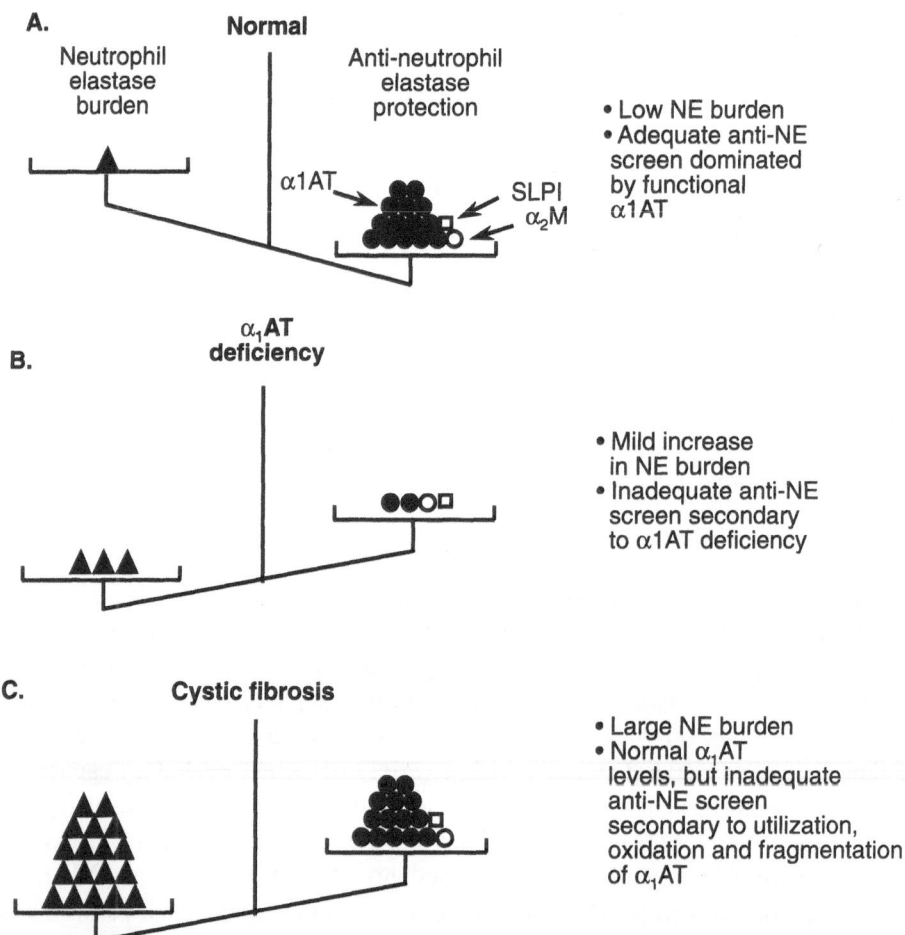

Fig. 1. Neutrophil elastase (NE) - anti-neutrophil elastase (anti-NE) balance in the normal lung and in the common pulmonary hereditary disorders. A. Normal. B. α1-antitrypsin (α1AT) deficiency. C. Cystic fibrosis. In normals, α1AT dominates the anti-NE screen. Other anti-neutrophil elastases present include the secretory leucoprotease inhibitor (SLPI) and α2-macroglobulin (α2M); together SLPI and α2M contribute <10% of the total anti-NE screen of the normal lung.

where it is available to diffuse across tissues. The major function of α1AT is to inhibit NE. The two molecules have a great avidity and they form a tight, non-covalent complex rendering the NE unable to function. In the normal lung, >95% of the α1AT is functional, capable of inhibiting NE.

SLPI, a 12 kDa non-glycosylated 107 residue single chain, disulphide linked serine antiprotease, is synthesized and secreted by cells of mucosal surfaces, including the epithelium of the upper airways.[6,11] Like α1AT, SLPI is capable of inhibiting a variety of serine proteases, but is best at inhibiting NE. Although approximately two-thirds of the SLPI in the normal respiratory epithelial surface is nonfunctional, there is sufficient SLPI in the airways to provide the major anti-NE capacity of the epithelial surface of the upper respiratory tract.[12]

α2M is a large (720 kDa) glycosylated broad spectrum antiprotease.[6] Although α2M is present in plasma and is produced by a variety of cell types, including lung fibroblasts, it is unlikely that α2M plays a significant role in protecting the lung against NE. Although α2M is a competent inhibitor of NE, its size precludes its movement across tissues, and thus it likely serves only to protect the cells that produce it.

There are two dramatic examples of the consequences to the lung of having ineffective defences against neutrophil elastase: α1-antitrypsin deficiency and cystic fibrosis (CF). Both are common hereditary disorders and both are lethal. It is the purpose of this review to present an overview of the major role of NE in the pathogenesis of these disorders, and the current strategies being utilized to re-establish the anti-NE defences of the lung to prevent the derangement of respiratory tissues that characterizes each disorder. For detailed references regarding these topics, several recent reviews are available.[2,7-10,13]

α1-Antitrypsin Deficiency

α1-antitrypsin deficiency is a hereditary disorder characterized by serum levels of α1AT less than 11μM, a high risk for the development of emphysema by the third or fourth decade, and a lesser risk for the development of liver disease, particularly in childhood.[7] The deficiency state is caused by mutations in the α1AT gene, a pleomorphic 12.2 kb, 7 exon gene on chromosome 14 at q31-31.2.[7,8]

Clinical manifestations

Emphysema is the most common manifestation of α1AT deficiency.[14,18] Respiratory symptoms develop at ages 25 to 40 years with gradually increasing dyspnoea, particularly with exercise, and occasionally a bronchospastic component. α1AT deficient individuals have a shortened life span of 10 to 15 years compared to the normal population and if there is history of cigarette smoking, life expectancy is further reduced.[9,10,14] Pulmonary function tests in α1AT deficiency are similar to those in other forms of emphysema, with severe limitation of airflow, reduction in diffusing capacity, increased total lung capacity and ratio of residual volume to total lung capacity. The typical chest radiograph demonstrates lung hyperinflation, symmetrical loss of parenchymal vascularity (primarily in the lower lung zones), bullae and a reduction in the cardiothoracic ratio. Computed chest tomography (CAT) is more sensitive in detecting "early" emphysematous change prior to pulmonary function abnormalities or symptomatic deterioration.[16]

As the disease progresses, a CAT scan typically shows a loss of parenchyma, large air spaces, and bullae (Fig. 2, compare panel B to panel A). Ventilation-perfusion lung scanning shows asymmetric delay in the washout of ^{133}Xenon from the lower zones, and symmetric loss of pulmonary arterial perfusion in the same regions.

Liver disease in association with α1AT deficiency occurs in two distinct age groups.[7,14] Approximately 10% of neonates with α1AT deficiency develop cholestasis with hepatitis, occasionally progressing to cirrhosis. In a small proportion of an older age group, usually >40 years, hepatitis and cirrhosis can develop, occasionally proceeding to liver failure.

There is also a clear association of α1AT deficiency with relapsing panniculitis.[15] Other disorders occasionally reported in association with α1AT deficiency include vasculitis, glomerulonephritis, uveitis and thyroiditis.

Pathogenesis

α1AT is the major lower respiratory tract inhibitor of NE, an omnivorous proteolytic enzyme stored in neutrophils.[1] NE is released following neutrophil activation (eg, associated with phagocytosis) or consequent to neutrophil damage and death. In normal individuals, the mild but chronic NE burden in the respiratory tract is balanced by an excess of α1AT. α1AT binds with NE in a pseudo-irreversible fashion, thus preventing NE from destroying the components of the extracellular matrix of the alveolar walls. In α1AT deficient individuals, there is a mild increase in the NE burden, an increase accentuated by cigarette smoking.[17] Coupled with the marked reduction in the anti-NE protective screen due to the α1AT deficiency state, this leaves the lung vulnerable to the unopposed actions of NE, with resultant destruction of alveoli leading to emphysema. The α1AT deficiency state is caused by mutations in the α1AT gene.[7-10] To date, 20 different α1AT

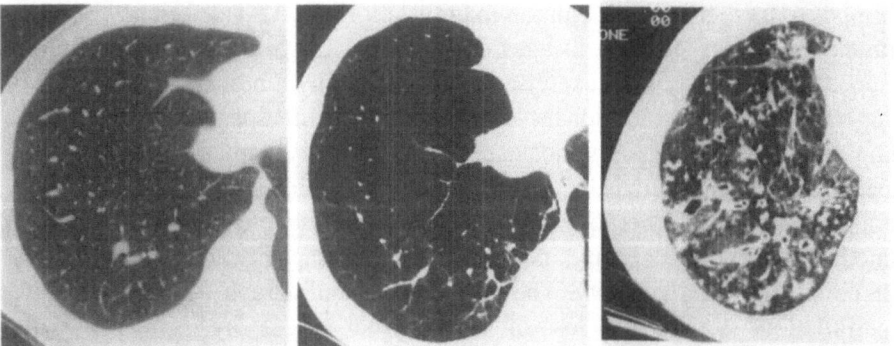

Fig. 2. Examples of computed axial tomography (CAT) scans of the thorax in a normal individual and in the common hereditary lung disorders. A. normal (left panel). B. α1-antitrypsin deficiency (middle panel). C. Cystic fibrosis (right panel). All scans are 1.5 mm cuts of the right lower lung zone.

alleles have been identified in association with α1AT (Table I). These can be classified into four groups depending on the types of mutations, including:

1. base substitutions:
2. in-frame deletion (eg, deletion of a single codon);
3. frame-shift caused by single or double base deletions or additions, resulting in a stop codon;
4. deletion of entire exons (see refs. 7 and 8 for details concerning these mutations and their consequences on expression of the α1AT gene).

Despite the many different alleles associated with α1AT deficiency, the Z mutation is, by far, the most common, representing >90% of the identified cases.[18] The Z mutation is common in the population of Caucasians of European descent with an allelic frequency of 1 to 2%.[7,8] As an example of the dominance of the Z mutations, in our experience with more than 500 individuals with α1AT deficiency, approximately 95% have been Z homozygotes.

The accentuation of the clinical consequences of the α1AT deficiency state by cigarette smoking is also explained by the fact that the α1AT molecule has a methionine residue at its active site (Met^{358}).[10]

This residue is vulnerable to free radicals such as those in cigarette smoke and those released by inflammatory cells recruited to the lungs in association with smoking. When the Met^{358} residue is oxidized, the association rate constant for inhibition of NE by α1AT is decreased 1000-fold or more, thus further depressing the lung defences against NE. Despite these explanations, the clinical presentation of α1AT deficiency is somewhat inhomogeneous, with some individuals developing severe emphysema at an early age with relatively little smoking history, while others (albeit rarely) retain reasonably good function throughout life despite heavy cigarette smoking.

One possible explanation for this diversity is that there might be genetic variations in the expression of the NE gene and/or storage of NE, and thus a relatively higher potential burden of NE within the lung. In this regard, higher levels of NE in neutrophils have been demonstrated in α1AT deficient patients with severe disease compared to those with minimal disease.[19]

The pathogenesis of α1AT deficiency related to the liver disease is less well understood. Interestingly, although a large variety of mutations of the α1AT gene cause α1AT deficiency, only two mutations, Z and M$_{malton}$ are clearly associated with an increased risk of liver disease.[7,8] Further, while individuals with the Null-Null phenotype have no serum α1AT, they do not exhibit liver disease, suggesting that it is not α1AT deficiency *per se* which is the culprit. It is hypothesized that the liver disease associated with Z and M$_{malton}$ alleles results from the accumulation of α1AT in hepatocytes, but the process by which accumulated α1AT damages the hepatocytes is not known, nor is it understood why only a minority of Z homozygotes develop liver disease.[7,14]

Table I. Classes of mutations of the α1-antitrypsin gene associated with α1-antitrypsin deficiency and emphysema[a]

Class [b]	Allele[c]
Base substitution	Z
	S
	$M_{heerlen}$
	$M_{mineral\ springs}$
	$M_{procida}$
	I
	P_{lowell}
	$W_{bethesda}$
	$Null_{bellingham}$
	$Null_{devon}$
	$Null_{ludwigshafen}$
	$Null_{iiyama}$
In-frame deletion	M_{malton}
	$M_{nichinan}$
Frameshift	$Null_{granite\ falls}$
	$Null_{matawa}$
	$Null_{hong\ kong}$
	$Null_{bolton}$
	$Null_{clayton}$
Exon deletion	$Null_{isola\ di\ procida}$

[a] List of α1AT alleles associated with α1AT deficiency for which a partial or complete sequence is known at the gene level; details regarding these mutations can be found in references 7-10.
[b] "Class" refers to the type of mutation at the gene level; **base substitution** = single base substitution; **in-frame deletion** = small deletion of one or more codons of coding exons that maintains reading frame; **frameshift** = single or double base deletion or insertion causing a frameshift and downstream stop codon; **exon deletion** = deletion of large portion of the α1AT gene including coding exons.
[c] Letters refer to position of migration on isoelectric focusing gel analysis of serum; "null" = no detectable α1AT in serum attributable to that allele (subscripts refer to birthplace of index case).

Cystic Fibrosis

Cystic fibrosis (CF), a lethal hereditary disorder manifest in organs with exocrine glands, is caused by mutations of the CF gene, a 27 exon, 250 kb segment of chromosome 7 at q31[13,20-22] (Fig. 3A).

The incidence of the disease varies among different ethnic group.[5]

A.

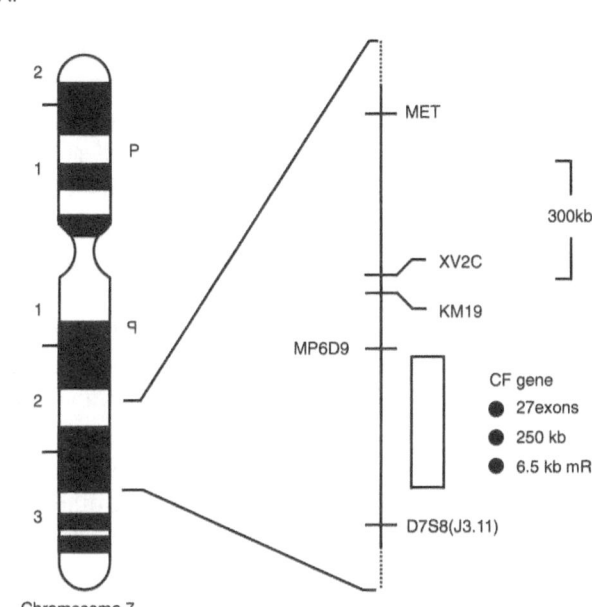

Chromosome 7

Fig. 3. Molecular basis of cystic fibrosis (CF).

A. Chromosomal localization of the CF gene. The CF gene is located on the q arm of chromosome 7 near the genomic markers MET, XV2C, and KM19, and between the markers MP6D9 and D7S8 (V3.11).[20-22] The CF gene is comprised of 27 exons spread over 250 kilobases (kb) and codes for mRNA transcripts of 6.5 kb.

B. Consequences of mutations of the CF gene on Cl⁻ secretion at the apical surface of epithelial cells. The putative protein product of the CF gene (referred to as the "CF transmembrane conductance regulator" or "CFTR") is 1480 amino acids in length with a predicted molecular weight of 168 kDa. It is likely, but not yet proven, that CFTR is part of regulatory units that modulate Cl⁻ channel(s). Mutations of the CF gene are associated with an inability of affected epithelial cells to respond to signals to increase Cl⁻ secretion.

B.

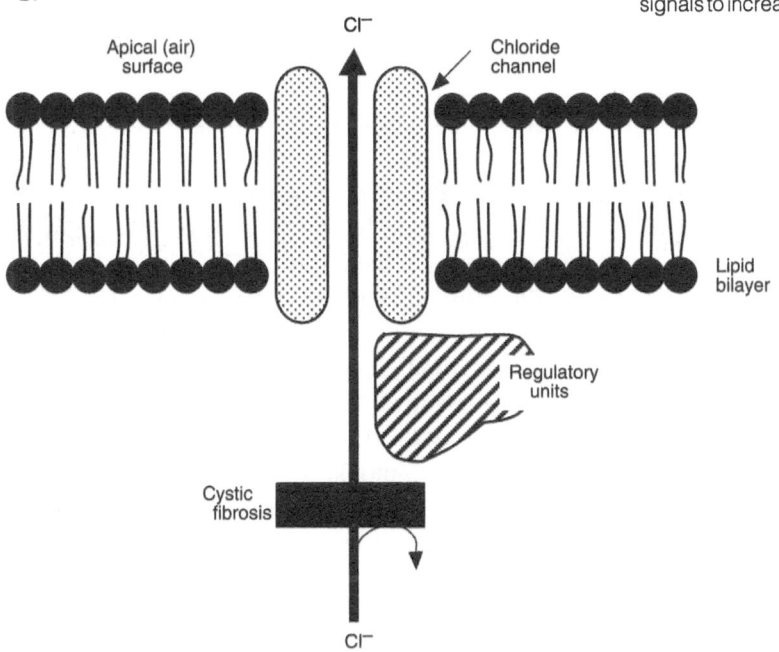

The highest incidence is among Caucasians, with the disease occurring in 1 in every 2000 births (allelic frequency 0.022).

The most common mutation of the CF gene associated with cystic fibrosis is a triple base deletion resulting in loss of phenylalanine at residue 508 of the predicted 1480 amino acid protein. Among North American Caucasians, approximately 46% of individuals with CF are homozygous for the ΔPhe^{508} mutation, 44% are compound heterozygous for this and other mutations, and 10% carry two other CF gene mutations.[22]

Although the function of the gene product (referred to as the "cystic fibrosis transmembrane conductance regulator" or "CFTR") is unknown, evidence from a variety of sources suggests that mutations of the CF gene result in abnormalities of ion transport across the respiratory epithelium, likely involving abnormalities in modulation of Cl^-, and perhaps Na^+, channel function in apical membranes.[23]

Clinical manifestations

The major clinical manifestations of CF are in the lung and the gastrointestinal tract.[13] In most cases, the respiratory problems dominate, and are usually the cause of death.

Typically, individuals with CF have a persistent cough productive of tenacious purulent sputum, haemoptysis, and recurrent respiratory infections. There is progressive derangement of the airways and the lower respiratory tract, including lung destruction. Eventually, respiratory failure develops.

Radiographically, the earliest change is hyperinflation followed by peribronchial cuffing, mucus impaction, bronchiectasis and peripheral round densities. A CAT scan of the chest shows cavities, mucus impaction, bullae and fibrotic changes (Fig. 2, compare panel C to panel A).

Lung function tests show increased airway resistance, diminished flow rates, gas trapping and a reduced diffusing capacity.

Airway reactivity is common. Periods of relative respiratory stability alternate with exacerbations, usually triggered by acute bacterial, viral or mycoplasmal infections. The airways of CF patients are colonized with bacteria such as *Staphylococcus aureus*, *Haemophilus influenzae* and *Pseudomonas aeruginosa* at an early stage.

Once established, these infections are rarely, if ever, eradicated. Exacerbations characteristically occur with increasing frequency with age, and intensive antibiotic therapy and mechanical assistance with clearance of mucus are usually required to control exacerbations of lung symptoms. As the disease progresses, there is limitation of activity, with a typical sequence of terminal events including hypoxaemia, pulmonary hypertension, respiratory failure, cor pulmonale and death usually by the third decade.

The gastrointestinal disturbances associated with CF begin in early childhood with failure to thrive and steatorrhoea due to exocrine pancreatic insufficiency.

In about 10% of newborn children with CF, meconium ileus occurs. The gastrointestinal abnormalities remain important features of disease throughout the life of the individual, but they are rarely life threatening because of therapy with oral pancreatic enzymes and vitamins. Approximately 30% of patients with CF have glucose intolerance and clinically significant diabetes mellitus occurs in 1 to 2% of all individuals with CF.

The incidence of diabetes increases with age; in patients over 25 years of age, diabetes occurs in 13%. Hepatic abnormalities occur in 20 to 50% of cases, but only 5% develop cirrhosis and 2% have clinically significant disease.

Pathogenesis of the lung derangements

It is not known how mutations in the CF gene cause the respiratory abnormalities observed in cystic fibrosis, but it is assumed to be directly related to abnormalities in expression of the CF gene in the respiratory epithelium (Fig. 4).

In this regard, it is likely that epithelial ion channel abnormalities that result from mutations of the CF gene are responsible for the abnormal mucus, bacterial colonization and inflammation that characterize the disease. How this occurs is not known, but it is probable that the inflammation is secondary to the chronic infection. The inflammation is dominated by neutrophils, and patients with CF have a very large burden of neutrophil derived proteases on the respiratory epithelial surface.[24,25] Although NE is likely not the only culprit mechanism causing lung derangements in CF, the chronic NE burden on the epithelial surface clearly plays a major role in this process. In this regard, not only can NE cause lung destruction (i.e., as in α1AT deficiency, see above), but it can directly injure epithelial cells and can interfere with defence capacities by reducing ciliary beat frequency, deranging the secretion of mucus glycoproteins, cleaving complement components and immunoglobulin, and interfering with the ability of neutrophils to kill *Pseudomonas.* [1,4,5]

The presence of significant quantities of NE, especially in combination with other inflammatory mediators, including myeloperoxidase and oxidant molecules also present on the CF epithelial surface, creates a highly toxic milieu which overwhelms the anti-NE protective screen of the respiratory epithelial surface. Consistent with this concept, evaluation of the α1AT and SLPI molecules in ELF of individuals with CF demonstrate they have been inactivated by proteolytic cleavage, oxidized, or have formed complexes with NE.

Thus, individuals with CF have a chronic burden of free NE on the respiratory epithelial surface, with consequent chronic derangement of airways and parenchyma.

178

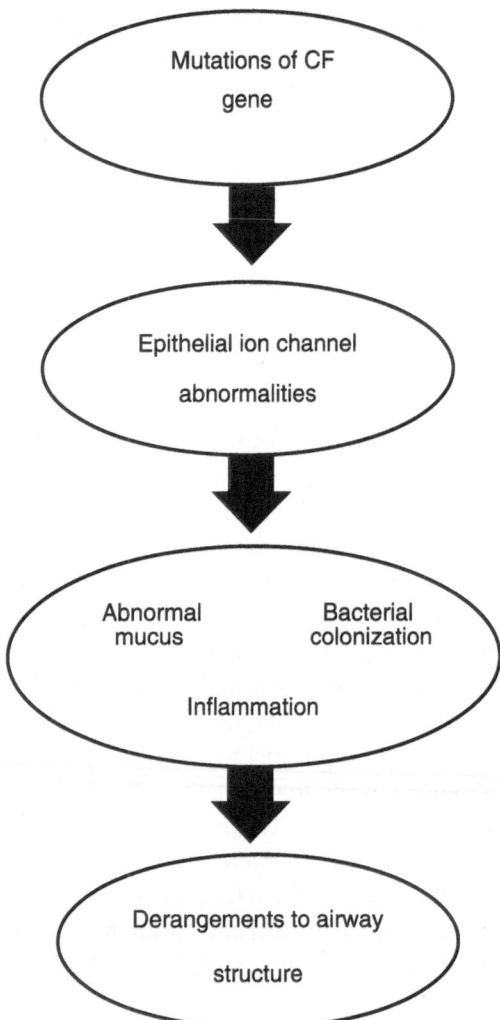

Fig. 4. Pathogenesis of the derangements to airway structures associated with mutations of the cystic fibrosis (CF) gene. Mutations of the gene cause abnormalities in the function of epithelial ion channels. Consequently, by mechanisms as yet not understood, there are abnormalities in mucus, bacterial colonization and inflammation. Together, these processes (particularly the inflammation) cause the derangements to airway structures.

Therapeutic Approaches to Prevent Lung Derangement in Hereditary Lung Disorders

Based on the knowledge of the pathogenesis of α1AT deficiency and cystic fibrosis, a rational therapeutic strategy for both disorders is to augment the anti-NE protective screen of the respiratory tissues. For α1AT deficiency, such an approach should be curative (i.e., prevent further loss of alveoli), since there is overwhelming evidence for the association of the α1AT deficiency state and emphysema. For cystic fibrosis, there are likely processes other than NE that are responsible for deranging the lung, but all available evidence suggests that NE plays a major role, and thus it is logical to consider therapies to suppress the respiratory epithelial NE burden in this disorder. In general, independent of the disease, three categories of therapeutic approaches are theoretically possible to augment the anti-NE protective screen of respiratory tissues, including:

1. augmentation of liver α1AT synthesis and/or secretion;
2. direct augmentation of lung anti-neutrophil elastase protection;
3. gene therapy (Table II).

Table II. Possible therapeutic approaches to augmenting the anti-neutrophil elastase protective screen of respiratory tissues.[a]

Augmentation of liver α1-antitrypsin synthesis/secretion
    ° Liver transplantation
    ° Impeded androgens
    ° Oestrogen analogues

Direct augmentation of lung anti-neutrophil elastase protection
    ° Plasma α1-antitrypsin
    ° Recombinant α1-antitrypsin
    ° Recombinant secretory leucoprotease inhibitor
    ° Low molecular weight chemical inhibitors of neutrophil elastase

Gene therapy

---

a. See text and references 2,7-10 for review

Augmentation of liver α 1-antitrypsin synthesis and/or secretion

Since the liver is the major site of α1AT for the body, one possible approach to augmenting the lung anti-NE screen would be to increase the amount of α1AT secreted daily by the liver. This strategy would clearly be useful in α1AT deficiency, as the major problem in this disorder is the deficiency state *per se*. It is unlikely to be useful in CF, as the plasma α1AT levels are normal, and even doubling plasma levels with the intravenous administration of purified plasma α1AT is insufficient to chronically suppress the NE burden on the respiratory epithelium in these individuals (see below).

Liver transplantation

Since the liver is the major site of α1AT production, theoretically liver transplantation would be the ideal augmentation therapy for α1AT deficiency, as it would replace Z homozygous hepatocytes with M homozygous hepatocytes. In this regard, in liver transplants performed in Z homozygotes with α1AT deficiency, conversion of serum α1AT to the α1AT phenotype of the transplanted liver occurs together with normalization of serum α1AT levels.[26] Liver transplantation has been used successfully to treat the liver diseases in α1AT deficient children and is the second most common indication for liver transplantation in children, with 5 year survival rates exceeding 70%.[27] The long term survival for liver transplantation in the adult group with α1AT deficiency and liver disease is somewhat lower at 60%.[27] However, despite being the definitive therapy for α1AT deficiency, liver transplantation has a number of important drawbacks, including the lack of donor organs, the need for long term immunosuppressive therapy, and the significant mortality rate. The latter is the most important; while acceptable for children or adults with progressive severe liver disease, the present rate is unacceptable in patients for a slowly progressive pulmonary disease.

Liver upregulation of α1AT production and/or secretion

Various attempts have been made to increase the endogenous production and/or release of α1AT by the liver in α1AT deficiency. Capitalizing on the fact that plasma α1AT levels are elevated in fever, trauma, shock and pregnancy, patients with α1AT deficiency have been treated in short term studies with typhoid vaccine and oestrogen-progesterone combinations.[9,10,14] However no significant increases in α1AT levels have been observed. Another strategy includes the use of danazol, a derivative of the synthetic steroid 17 α-ethinyl testosterone, with properties similar to testosterone but lacking its androgenic properties.[28] The rationale behind this strategy is based on the knowledge that danazol increases serum concentrations of C1-esterase in hereditary angio-oedema, another disorder characterized by serum deficiency of a hepatocyte-produced antiprotease. However while providing a modest increase in α1AT serum levels in some α1AT deficient individuals, the increases associated with danazol are insufficient to provide protective levels in serum and respiratory ELF.

Another strategy involves the use of tamoxifen, an agent which binds to intracytoplasmic oestrogen receptors and thus may mimic the increases in serum α1AT levels that normally occur in pregnancy. However, while trials with tamoxifen have shown increases in serum α1AT above the protective levels in individuals with α1AT phenotypes SZ and PZ, when administered to Z homozygous individuals, tamoxifen does not result in increases in α1AT levels sufficient to re-establish the normal anti-NE protection of the lung.[29]

Direct augmentation of lung anti-neutrophil elastase protection

The direct approach to augmenting the lung anti-NE protection is to administer molecules with anti-NE properties, either systemically or directly to the lung. This strategy is relevant to both α1AT deficiency and CF, although the clinical experience has been mostly with α1AT deficiency. The molecules that have been used for this purpose in clinical and/or experimental animal studies include: plasma α1AT, recombinant α1AT, recombinant SLPI and low molecular weight chemical inhibitors of NE.[2,11,30,31]

Plasma α1-antitrypsin

The simplest approach to therapy for α1AT deficiency is to augment serum and lower respiratory tract α1AT with exogenous plasma α1AT. Early studies using a crude preparation of human α1AT purified from human plasma and given intravenously once weekly for four weeks showed this to be practicable with significant increases in serum and ELF α1AT.[32] Following these studies, methods were developed to purify α1AT from pooled human plasma on a commercial scale, and for several years the Pulmonary Branch, NHLBI, conducted trials of α1AT augmentation therapy for α1AT deficiency using an α1AT preparation partially purified from pooled human plasma of normal donors.[30] This preparation is heat treated to lower the risk of plasma-borne viral disease. It is 80 % pure, the remainder being composed of various plasma proteins. The α1AT in the preparation is 70 to 80 % active, and the active molecules have a normal association rate constant for NE.

Because the serum half-life of α1AT is 4 to 5 days, weekly infusions of α1AT of doses of 60 mg/kg are sufficient to maintain serum levels of α1AT above 11 μM, a level sufficient to provide adequate anti-NE protection for the lung.[30] Not only are serum α1AT levels increased, but so are levels in respiratory ELF, so that even at nadir levels six days after intravenous administration of α1AT, ELF levels are increased sufficiently. Importantly there is a concomitant increase in anti-NE capacity in serum and ELF, thus demonstrating the biochemical efficacy of α1AT augmentation therapy for α1AT deficiency. Further studies have demonstrated that dosages of 250 mg/kg given intravenously once monthly produced similarly efficacious results.[33]

An alternative strategy to use plasma α1AT to treat α1AT deficiency is to administer the purified plasma αAT by aerosolization directly to the lower respiratory tract.[34]

This approach has been proven to be safe and dosages of 100 mg α1AT aerosolized every 12 hours for 7 days result in significant elevations in α1AT levels in ELF well above the protective threshold level, with a concomitant normalization of ELF anti-NE defences. The problem with this approach is that while respiratory epithelial anti-NE augmentation can be achieved, it is not clear whether enough α1AT diffuses across the epithelium to provide sufficient protection to the alveolar interstitium. Experiments in sheep with cannulation of the thoracic lymph duct showed that aerosolized α1AT diffuses across the respiratory epithelium entering lung interstitial lymph, and reaches the systemic circulation[34], but this is difficult to prove in man. Whether this therapy will prove to be as effective in preventing parenchymal lung destruction in α1AT deficiency as intravenous α1AT augmentation therapy remains to be seen, but it is easier on the patients, and likely will require less drug. In this regard, with doses of 100 mg twice daily, α1AT accumulates on the alveolar surface, perhaps providing a natural "depot" for the drug i.e., perhaps after an initial treatment period, it may be possible to reduce the therapy to once daily.[34]

Lacking evidence that NE plays any positive role in CF and with the background of the understanding of the role of NE in the pathogenesis of lung damage in this disorder, it is rational to consider plasma α1AT for the therapy of CF. With the example of α1AT augmentation therapy for α1AT deficient patients, initial studies used the intravenous route for administration of plasma α1AT.[25] However, while α1AT given in this manner to CF patients is capable of suppressing the NE burden on the respiratory epithelium when given in large doses up to 120 mg/kg weekly, the effect is transient. In this regard, one week following administration, active NE is still present on the epithelial surface suggesting that, while intravenous administration of large amounts of α1AT is capable of temporarily suppressing the NE burden in CF, it is unable to do so on a sufficiently chronic basis to act as a viable therapy.[25]

While aerosol α1AT therapy may or may not be efficacious for α1AT deficiency, it should be efficacious for cystic fibrosis (CF), a disorder where the NE burden is confined to the respiratory epithelial surface. In this regard, α1AT given by aerosol twice daily to individuals with CF increases ELF levels of α1AT, so that 12 hours after aerosolization, ELF α1AT levels are elevated, an effect maintained through the period of therapy.[24] All CF patients in this study had active NE in ELF pre-therapy, but if ELF α1AT levels increased to ≥8 μM post-therapy, the active NE burden in ELF was totally suppressed with resultant return of normal anti-NE capacity in ELF. Thus, α1AT aerosol therapy can suppress the respiratory NE burden in CF, likely preventing the direct toxic effect of NE on the lung.

The suppression of NE in CF also should prevent its deleterious effects on local host defences, with important implications for the intractable bacterial colonization of the airways in CF. In this context, NE is capable of cleaving the C3b receptor on neutrophils and as a consequence inhibits the ability of neutrophils to kill *Pseudomonas* effectively.[4] This host defence "defect" is observed in neutrophils exposed to CF epithelial lining fluid containing NE, with resulting decrement in the ability of the neutrophils to kill *Pseudomonas*.[35] Importantly, following therapy with aerosolized α1AT and subsequent suppression of the NE burden in ELF, the inhibitory ability of CF ELF on *Pseudomonas* killing by neutrophils can be circumvented. These observations raise the possibility not only of preventing further lung destruction in CF with α1AT aerosol augmentation therapy, but also of restoring host defences with subsequent eradication of colonizing organisms. Such a hypothesis clearly requires further long term studies involving large numbers of patients over a long time interval.

Recombinant α1-antitrypsin

One possible problem with using plasma purified α1AT for the therapy of α1AT deficiency and CF is the limited available supply. There are 20,000 to 40,000 Z homozygotes in the United States alone; therapy of all of these individuals would require 4000 to 8000 kg of purified α1AT yearly.[36] There are even more patients with CF, putting further demands on the available supply as this therapy is adopted for treating CF on a widespread basis. Secondly, while therapy with plasma purified α1AT has been shown to be safe, it is a plasma product with theoretical risks of viral infection and the adverse reactions typical of chronic administration of plasma products.[30]

An alternative source of α1AT is that produced by recombinant DNA technology (rα1AT). Plasmids containing human α1AT cDNA have been used to transform *Escherichia coli* or yeast, and the purified α1AT product rα1AT is capable of inhibiting human NE in a similar fashion to human α1AT.[31] The rα1AT produced in yeast has a molecular mass of 45 kDa, no carbohydrates and is identical in sequence to normal α1AT except for an additional N-terminal acetylmethionine. Unfortunately, when given intravenously to primates, it disappears rapidly from blood, being barely detectable at 24 hours, with 38% of it being excreted in urine within 3 hours (unlike plasma α1AT, where none appears in urine following intravenous administration).[36]

This phenomenon obviates chronic intravenous administration of rα1AT to humans, but as an alternative, direct targeting of the lower respiratory tract by aerosolization of rα1AT may be used. As an example, after single aerosol administration of 200 mg of rα1AT to α1AT deficient individuals, ELF α1AT levels and anti-NE capacity were augmented in proportion to the dose of rα1AT administered, with an anti-NE capacity in ELF increased 40-fold over pre-therapy levels 4 hours following the aerosol.[31] rα1AT was also detectable in serum after aerosolization, indicating that the lower respiratory tract epithelium is permeable to rα1AT and thus, rα1AT has access to the interstitium. However, whether the levels of rα1AT in the human interstitium are sufficient to protect against lung destruction in α1AT deficiency will require further evaluation.

Recombinant secretory leucoprotease inhibitor

Another candidate for augmenting anti-NE protection in the lung is recombinant secretory leucoprotease inhibitor (rSLPI), a molecule identical to the naturally occurring 12 kDa non-glycosylated, disulphide linked molecule.[11] Animal studies have shown that rSLPI augments anti-NE capacity in ELF following aerosolization and once or twice daily aerosols are sufficient to maintain what should be protective, at least for the epithelium.[11] Possible advantages of rSLPI over α1AT or rα1AT include: [11]

1. it is acid-stable and may thus remain functional in the decreased pH surrounding neutrophils during the metabolic burst;

2. it has a pI>9, similar to that of NE, and may thus track and bind to tissue sites favoured by NE such as elastin, and there is evidence that SLPI can inhibit elastin-bound NE; and

3. the lower molecular mass of rSLPI may allow it to get to tissue sites "covered" by neutrophils, thus inhibiting NE in areas that cannot be reached by the larger α1AT molecule.

It has been proven, however, that rSLPI will reach the interstitium in levels to treat α1AT deficiency. In this regard, while rSLPI has a molecular mass of 27% that of rα1AT, simultaneous aerosolization to sheep of both molecules demonstrated that, on a mole for mole basis, significantly more rα1AT than rSLPI molecules were detectable in lung lymph.[11] This could be explained by compartmentalization of SLPI to the epithelial surface, by adsorption of rSLPI onto epithelial cells or other molecules of ELF, or by binding of rSLPI to molecules in the interstitium after passing through the epithelium. However, even if aerosolized rSLPI is not as good a candidate as α1AT for defending the interstitium of the lung against NE (and thus possibly not as effective for α1AT deficiency), it clearly will function to augment the anti-NE protective screen of the pulmonary epithelial surface, and thus be effective in cystic fibrosis.

Low molecular weight chemical inhibitors of neutrophil elastase

A number of low molecular weight elastase inhibitors have been shown to be effective in preventing or attenuating NE mediated pulmonary damage in animal models.[2] The most promising of these include peptide chloromethyl ketones, peptide aldehydes, peptide boronic acid, β lactams and eglin C. However, the data to date on these molecules is derived from animal experimentation and the question of toxicity and efficacy in humans remains to be tested.

Gene therapy

Several strategies have been devised to use the methods of gene transfer to augment the α1AT levels in the lung, making gene therapy an attractive approach to treating α1AT deficiency and cystic fibrosis. In addition to the question of safety of gene therapy, the major theoretical hurdle to using this approach to treat humans is the problem of getting the target cells to produce sufficient amounts of α1AT to provide sufficient anti-NE protection to the lung. Two approaches are currently under investigation; both are theoretically applicable to treating α1AT deficiency and cystic fibrosis.

First, T-lymphocytes can be modified to produce α1AT by inserting an active promoter plus the α1AT cDNA into the T-cell genome using a retroviral vector.[37] T-cells can be easily grown *in vitro*, and expanded clones producing α1AT could be administered intravenously. Alternatively, the modified T-cells could be directly transplanted to the epithelial surface of the lung, to provide "local" production of α1AT.

Second, a promoter-α1AT cDNA combination could be transferred directly to the respiratory epithelium, thus providing α1AT secreted by the epithelial cells. The feasibility of this approach has been demonstrated by using a modified adenoviral vector to directly modify rat lung epithelial cells *in vivo*.[38]

References

1. Hubbard R., Brantly M., Crystal R.G.: Proteases. In: Crystal R.G., West J.B., (editors-in chief) Barnes P., Weibel E.R., Cherniack N.S. (associate editors). *The Lung. Scientific Foundations.* New York, Raven Press, 1991; 1763-1774
2. Hubbard R., Crystal R.G.: Susceptibility of the lung to proteolytic injury. In: Crystal R.G., West J.B. (editors-in-chief), Barnes P., Weibel E.R, Cherniack N.S. (associate editors). *The Lung. Scientific Foundations.* New York, Raven Press, 1991; 2059-2072
3. Lucey E.C., Stone P.J., Christensen T.G., Breuer R., Snider G.L.: An 18-month study of the effects on hamster lungs of intratracheally administered human neutrophil elastase. Exp. Lung. Res. 1988; 14: 671-686
4. Berger M., Soërensen R.U., Tosi M.F., Dearborn D.G., Döring G.: Complement receptor expression on neutrophils at an inflammatory site, the pseudomonas-infected lung in cystic fibrosis. J. Clin. Invest. 1989; 84: 1302-1313
5. Fick R.B., Naegel G.P., Squier S., Wood R.E., Gee J.B.L., Reynolds H.Y.: Proteins of the cystic fibrosis respiratory tract: Fragmented immunoglobulin G opsonic antibody causing defective opsonophagocytosis. J. Clin. Invest. 1984; 74: 236-248
6. Hubbard R., Crystal R.G.: Antiproteases. In: Crystal R.G., West J.B. (editors-in-chief), Barnes P., Weibel E.R., Cherniack N.S. (associate editors). *The Lung. Scientific Foundations.* New York, Raven Press, 1991; 1775-1788
7. Crystal R.G.: α1-antitrypsin deficiency, emphysema, and liver disease: Genetic basis and strategies for therapy. J. Clin. Invest. 1990; 85: 1343-1352
8. Crystal R.G.: The α1-antitrypsin gene and its deficiency states. Trends Genet 1989; 5: 411-417
9. Crystal R.G., Brantly M.L., Hubbard R.C., Curiel D.T., States D.J., Holmes M.D.: The α1-antitrypsin gene and its mutations: Clinical consequences and strategies for therapy. Chest 1989; 95: 196-208
10. Brantly M., Nukiwa T., Crystal R.G.: Molecular basis of α1-antitrypsin deficiency. Am. J. Med. 1988; 84: 13-31
11. Vogelmeier C., Buhl R., Hoyt R.F., Wilson E., Fells G.A., Hubbard R.C., Schnebli H-P, Thompson R.C., Crystal R.G.: Aerosolisation of recombinant secretory leucoprotease inhibitor to augment the anti-neutrophil elastase protection of pulmonary epithelium. J. Appl. Physiol. 1990; 65: 1843-1848
12. Vogelmeier C., Hubbard R., Geiger R., Fritz H., Crystal R.G.: Secretory leucoprotease inhibitor levels in normal upper and lower respiratory tract epithelial lining fluid of normals. Am. Rev.

Respir. Dis. 1989; 139: A200

13. Boat T.F., Welsh M.J., Beaudet A.L.: Cystic Fibrosis. In Scriver C.R., Beaudet A.L., Sly W.S., Valle D. (Eds.) *The metabolic basis of inherited disease*. 6th ed. New York, McGraw-Hill 1989; 2649-2680

14. Cox D.W.: α1-antitrypsin deficiency. In: Scriver C.R., Beaudet A.L., Sly W.S., Valle D. (Eds.) *The metabolic basis of inherited disease*. 6th ed. New York, McGraw-Hill 1989; 2409-2437

15. McElvaney N.G., Feuerstein I., Simon T.R., Hubbard R.C., Crystal R.G.: Comparison of the relative sensitivity of routine pulmonary function tests, scintigraphy, and computed axial tomography in detecting "early" lung abnormalities associated with α1-antitrypsin deficiency. Am. Rev. Respir. Dis. 1989; 139: A122

16. Pittelkow M.R., Smith K.C., Su D.W.P.: α-1-antitrypsin deficiency and panniculitis. Am. J. Med. 1988; 84: 80-85

17. Hunninghake G.W., Crystal R.G.: Cigarette smoking and lung destruction: accumulation of neutrophils in the lungs of cigarette smokers. Am. Rev. Respir. Dis. 1983; 128: 833-838

18. Brantly M.L., Paul L.D., Miller B.H., Falk R.T., Wu M., Crystal R.G.: Clinical features and natural history of the destructive lung disease associated with α-1-antitrypsin deficiency of adults with pulmonary symptoms. Am. Rev. Respir. Dis. 1988; 138: 327-336

19. Hubbard R.C., McElvaney N.G., Crystal R.G.: Amount of neutrophil elastase carried by neutrophils may modulate the extent of emphysema in α-1-antitrypsin deficiency. Am. Rev. Respir. Dis. 1990; 141: A683

20. Rommens J.M., Iannuzzi M.C., Kerem B-S., Drumm M.L., Melmer G., Dean M., Rozmahel R., Cole J., Kennedy D., Hidaka N., Zsiga M., Buchwald M., Riordan J.R., Tsui L-C., Collins F.S.: Identification of the cystic fibrosis gene: Chromosome walking and jumping. Science 1989; 245: 1059-1065

21. Riordan J.R., Rommens J.M., Kerem B-S, Alon N., Rozmahel R., Grzelczak Z., Zielenski J., Lok S., Plavsic N., Chou J-L., Drumm M.L., Iannuzzi M.C., Collins F., Tsui L-C.: Identification of the cystic fibrosis gene: Cloning and characterization of complementary DNA. Science 1989; 245: 1066-1073

22. Kerem B-S., Rommens J.M., Buchanan J.A., Markiewicz D., Cox T.K., Chakravarti A., Buchwald M., Tsui L-C.: Identification of the cystic fibrosis gene: Genetic analysis. Science 1989; 245:1073-1080

23. Welsh M.J., Fick R.B.: Cystic fibrosis. J. Clin. Invest. 1987; 80:1523-1526

24. McElvaney N.G., Hubbard R.C., Fells G.A., Chernick M.C., Caplan D.B., Crystal R.G.: Aerosolization of α1-antitrypsin to establish a functional anti neutrophil elastase defense of the respiratory epithelium. Clin. Res. 1990; 38:485

25. McElvaney N.G., Hubbard R.C., Fells G.A., Healy J., Chernick M.C., Crystal R.G.: Intravenous α1-antitrypsin to reestablish anti-neutrophil elastase defenses of the pulmonary epithelial surface in cystic fibrosis. Am. Rev. Respir. Dis. 1990; 141:A83

26. Esquivel C.O., Vicente E., Van Thiel D., Gordon R., Marsh W., Makowka L., Koneru B., Iwatsuki S., Madrigal M., Delgado Milan M.A., et al.: Orthoptic liver transplantation for α1-antitrypsin deficiency: an experience in 29 children and 10 adults. Transplant. Proc. 1987; 19: 3798-3802

27. Esquivel C.O., Marino I.R., Fioravanti V., Van Thiel D.H.: Liver transplantation for metabolic disease of the liver. In: Gastroenterology Clinics of North America 1988; 17 (1): 167-177

28. Wewers M.D.Gadek J.E., Keogh B.A., Fells G.A., Crystal R.G.: Evaluation of danazol therapy for patients with PiZZ α-1-antitrypsin deficiency. Am. Rev. Respir. Dis. 1986; 134: 476-480

29. Wewers M.D., Brantly M.L., Casolaro M.A., Crystal R.G.: Evaluation of tamoxifen as a therapy to augment α 1-antitrypsin levels in Z homozygous α1-antitrypsin deficient individuals. Am. Rev. Respir. Dis. 1987; 135: 401-402

30. Wewers M.D., Casolaro M.A., Sellers S.E., Swayze S.C., McPhaul K.M., Wittes J.T., Crystal R.G.: Replacement therapy for α1-antitrypsin deficiency associated with emphysema. N. Engl. J. Med. 1987; 316:1055-1062

31. Hubbard R.C., McElvaney N.G., Sellers S.E., Healy J.T., Czerski D.B., Crystal R.G.: Recombinant DNA-produced α1-antitrypsin administered by aerosol augments lower respiratory tract anti-neutrophil elastase defenses in individuals with α1-antitrypsin deficiency. J. Clin. Invest. 1989; 84: 1349-1354

32. Gadek J.E., Klein H.G., Holland P.V., Crystal R.G.: Replacement therapy of α1-antitrypsin deficiency. J. Clin. Invest. 1981; 68: 1158-1165

33. Hubbard R., Sellers S., Czerski D., Stevens L., Crystal R.G.: Biochemical efficacy and safety of monthly augmentation therapy for α1-antitrypsin deficiency. JAMA 1988; 260: 1259-1264

34. Hubbard R.C., Brantly M.L., Sellers S.E., Mitchell M.E., Crystal R.G.: Delivery of proteins for therapeutic purposes by aerosolisation: direct augmentation of anti-neutrophil elastase defenses of the lower respiratory tract in α1-antitrypsin deficiency with an aerosol of α1-antitrypsin. Ann. Intern. Med. 1989; 111: 206-2129

35. McElvaney N.G., Hubbard R.C., Birrer P., Chernick M.S., Caplan D.B., Frank M.M., Crystal R.G.: Aerosol administration of α1-antitrypsin to suppress the burden of active neutrophil elastase on the respiratory epithelial surface in cystic fibrosis. *Submitted*

36. Casolaro M.A., Fells G., Wewers M., Pierce J.E., Ogushi F., Hubbard R.C., Sellers S., Forstrom J., Lyons D., Kawasaki G., Crystal R.G: Augmentation of lung antineutrophil elastase capacity with recombinant human α1-antitrypsin. J. Appl. Physiol. 1987; 63: 2015-2023

37. Curiel D., Stier L., Crystal R.G: Gene therapy for α1-antitrypsin deficiency using lymphocytes as vehicles for α1-antitrypsin delivery. Clin. Res. 1989; 37 (2): 578A

38. Crystal R.G, Rosenfeld M., Siegfreid W., Lantero S., Stratford-Perricaudet L., Dalmans W., Parivani A., Perricaudet M., LeCocq J-P.: New viral transfer systems. Ped. Pul. (in press)

14. Workshop Summary

J. Travis

Department of Biochemistry, University of Georgia, Athens, Georgia, USA

It has been almost 30 years since Laurell & Eriksson first demonstrated the relationship between α-1-proteinase inhibitor (α-1-PI) deficiency and the development of pulmonary emphysema.

Since that time considerable data has been generated which now shows that a) neutrophil elastase (HNE) is the primary culprit in the uncontrolloled degradation of lung connective tissue leading to emphysema, and b) besides genetic deficiencies resulting in abnormally low secretion of α-1-PI into the circulation, alternative mechanisms for reducing functional levels of normally secreted inhibitor (e.g., oxidation and/or proteolytic inactivation) also exist which may explain how "garden variety" emphysema occurs. That may be the bad news; however, the good news is far more encouraging in that several avenues for augmenting inhibitor levels in human tissue, such as infusion of native or recombinant forms of α-1-PI, other types of human inhibitors, synthetic inhibitors, or even gene therapy, show good promise for therapeutic use. Indeed, a plethora of information now exists which may result in drugs which will make it possible to aid patients at the earliest stages of emphysema and reduce the rate at which lung destruction occurs. Much of this is expanded upon in the review by Dr. Crystal and his group who have pioneered investigations involving the infusion of proteinase inhibitors for the reduction of HNE-induced lung connective tissue degradation. In fact, perhaps the only major barrier which still exists is with regard to detecting emphysema at its earliest stages, using biochemical parameters. Here, there is still plenty to be done.

This symposium describes in more detail some of the highlights which I have discussed very briefly above and will allude to in subsequent sections. For example, in the chapter by Dr. Barrett we see that HNE is not the only mammalian enzyme which can degrade elastin and collagen, cathepsins H and L, and metalloendopeptidases also playing a role in the degradation of these connective tissue components. Presumably such enzymes are involved in the local turnover of these proteins in both normal and inflammatory states, although Dr. Bieth has come to the conclusion that in centrilobular emphysema these may be just as important as HNE. The argument still remains, however, as to whether the concentration of these other enzymes, relative to HNE, is high enough to provide the damage which occurs during the development of this or any other kind of emphysema.

Dr. Stockley and Dr. Burnett also have questioned whether HNE is the only enzyme of importance in the development of emphysema. Their contention is that since HNE is so non-specific in its action on connective tissue proteins of the lung that other cells which accumulate in this organ (e.g., monocytes/macrophages) may also be capable of producing enzymes which can complement the activity of the neutrophil proteinases. Again, the concentration of these enzymes is considerably below that found in the neutrophil, and there still does not appear to be enough justification for giving the macrophage/monocyte system any major position in lung proteolysis, although a role in oxidative damage cannot be excluded. On the other hand, one must worry about the contribution of bacterial proteinases, such as *Pseudomonas aeruginosa* elastase, to the development of emphysema particularly since this organism grows so well in the lungs of cystic fibrosis patients.

Most individuals working in the emphysema field have chosen to attack the problem by investigating ways to stop elastin degradation. Dr. Davidson, however, has taken a different tack and examined the possibilities for increasing elastin synthesis.

This has required lengthy studies to examine what regulates elastin biosynthesis, one of the long term goals being to develop methods for increasing elastin synthesis by activating the gene(s) responsible for the production of this protein. Indeed, a long asked question, still not answered, revolves around the fact that the destruction of the integrity of the alveolus through proteolytic action is essentially irreversible.

Clearly, a continuation of the cell biology involved in the laying down of connective tissue matrices is of high importance.

The development of inhibitors against HNE has required a better understanding of the mechanism by which this enzyme functions.

Rather than looking at simple substrates, as most others have done, Dr. Baici has chosen to examine how this enzyme interacts with protein substrates, including elastin and plasma proteins.

Using a very nice kinetic system he has obtained data which suggests that there are major differences in the interaction of HNE with these two classes of proteins, being fast and reversible with soluble proteins but very slow and nearly irreversible with elastin. Thus, the development of inhibitors against HNE must take into account the difficulty of displacing HNE from the elastin surface, whereas this is not nearly the problem with soluble substrates. The size of the inhibitor is also of concern, the data obtained clearly showing that small molecules are much more able to compete with elastin for HNE, as previously shown also by Dr. Bieth's group.

The major approach to reducing proteinase levels in the lung, particularly HNE, has used the philosophy suggesting that increasing inhibitor levels in the tissues, exogenously, would be of most benefit. Thus, a number of studies have been performed to develop new kinds of inhibitors against the neutrophil enzyme. Dr. Powers' group has been at the forefront in the synthesis of such compounds which are of a considerably diverse structure. As pointed out by Dr. Powers, there is likely to be success in the use of at least two types of inhibitors, the peptide fluoroketones and the beta-lactams, primarily because of their greater stability in plasma. While this seems very exciting, a word of caution must still be expressed since it is not known, as yet, how many functions HNE has within the body.

How will the addition of small inhibitors affect neutrophil movement, phagocytosis, uptake of HNE-synthetic inhibitor complexes, etc.? These are questions which must be addressed prior to the general use of synthetic inhibitors.

It is more likely, then, that human HNE inhibitors will have the better chance of general success in reducing connective tissue damage associated with the development of emphysema. Dr. Kramps' group address this approach in their paper dealing with the secretory leucocyte proteinase inhibitor (SLPI). Their results are most exciting because the inhibitor is a) human and, therefore, non-antigenic, b) considerably larger than synthetic inhibitors and, therefore, less able to bind to HNE in areas where this enzyme may have considerable importance in maintaining homeostasis, c) inhibits HNE rapidly both in solution and when the enzyme is bound to elastin, and d) can be produced by recombinant DNA technology. Finally, e) SLPI appears to have all of the properties required for it to be given by aerosolization.

Here there is real hope for a solid candidate for therapy.

However, SLPI is not the only inhibitor which may prove useful in HNE therapy. As pointed out by Dr. Fritz there are several other choices. These include homologues of aprotinin from bovine mast cells, Kazal-type inhibitors such as the human secretory trypsin inhibitor and Kunitz-type inhibitors (e.g., the domain found in the amyloid precursor protein associated with Alzheimer's disease). Such proteins will have to be modified through genetic engineering in order to make them more efficient inhibitors of HNE, and this may cause antigenic reactions, particularly during long-term therapy. Nevertheless, they are intriguing alternatives to both α-1-PI, and synthetic inhibitors of HNE.

As pointed out earlier, one of the difficulties in evaluating whether drugs against HNE really work will depend not only on long-term clinical studies but also on whether one can measure reduction in lung connective tissue damage. Therefore, the quandary still exists as to how to measure changes in specific levels of elastin or collagen degradation products (e.g., desmosine) during such studies. As pointed out by Dr. Snider's group approaches to measure such changes have been fraught with failure. Almost certainly, as he adds, this is due to methodological errors, making it virtually impossible to measure changes in connective tissue protein turnover after inhibitor treatment and, indirectly, the ability to determine the efficacy of the drug being tested. If there were ever an area sorely in need of more good researches, in addition to those addressing the proteinase inhibitor aspects in emphysema research, it is here.

Since the primary problem in the development of emphysema is the lack of sufficient α-1-PI to regulate HNE being released by neutrophils, why not get the liver to make more inhibitor? This is essentially the approach taken by Dr. Woo's group who believe that hepatic gene therapy would be appropriate, particularly for those who are α-1-PI deficient. There is no doubt that this is the approach not for the distant future but certainly within the current decade. The data given here and also by others clearly supports the feasibility of increasing inhibitor levels in this manner and completely eliminates the criticisms associated with other types of inhibitor therapy.

In summary, great strides have been made in addressing ways to reduce connective tissue damage in the lung. Since the whole perspective of emphysema development should be properly described as a chronic inflammatory problem, it is clear that all of the studies, to date, are likely to be useful in the treatment of similar diseases such as rheumatoid arthritis, glomerulonephritis, gingivitis, etc. A similar situation is clearly evolving in the study of AIDS, where a plethora of knowledge is now accumulating which should be useful in understanding and treating other viral-induced diseases. To those working in the field of emphysema I can only say that in the not too distant future the work that has been done in this area will ultimately be recognized as fundamental to the understanding and treatment of most chronic inflammatory diseases.

Subject Index

A

α-1 protease inhibitor (α1-PI) 72, 103, 143
α_1-proteinase inhibitor (α_1-PI) 1, 33, 35, 42, 94
- - chemical properties of 72
α1-antichymotrypsin (α1-Achy) 4
α1-antiprotease therapy 21
α1-antitrypsin (α1-AT) 48, 159, 169
- deficiency 113, 160, 170, 171, 181
- synthesis 179
α2-macroglobulin (α-2M) 4, 169, 170
α-chymotrypsin 114
Active proteinase inhibitors 56
Acute bacterial pneumonia 58
Adsorption of elastase 92
- - on elastin and elastolytic reaction 92
Adult respiratory distress syndrome 123
Aerosolization 181
Aerosols 169
Alveolar walls 113
Alveolus 21
Alzheimer amyloid protein precursor 104, 106
Aminoalkylphosphonate diphenyl 128
Anti-inflammatory agents 8
Antiemphysema drugs 43
Antileucoprotease (ALP) 4, 52, 105, 114, 113
Antineutrophil elastase screen 113
Antioxidants 144
Antiproteases 143
Antiproteinase 47
Antithrombin III (AT III) 101, 104
Aprotinin 104
- homologues 106
Aspartic endopeptidases 30, 32
Atherosclerosis 123
Auranofin 8
Azurophil granule 58

B

Bacterial elastases 4
- proteinases 41